Thomas Henry Watson

Naval Architecture

A Manual on Laying-Off Iron, Steel, and Composite Vessels

Thomas Henry Watson

Naval Architecture
A Manual on Laying-Off Iron, Steel, and Composite Vessels

ISBN/EAN: 9783337180850

Printed in Europe, USA, Canada, Australia, Japan

Cover: Foto ©berggeist007 / pixelio.de

More available books at **www.hansebooks.com**

Naval Architecture:

A

MANUAL ON LAYING-OFF

IRON, STEEL, AND COMPOSITE VESSELS.

BY

THOMAS H. WATSON,

LECTURER ON NAVAL ARCHITECTURE AT THE DURHAM COLLEGE OF SCIENCE,
NEWCASTLE-UPON-TYNE;
MEMBER OF THE NORTH EAST COAST INSTITUTION OF
ENGINEERS AND SHIPBUILDERS.

WITH NUMEROUS ILLUSTRATIONS.

LONDON: LONGMANS, GREEN, & CO., 39, PATERNOSTER ROW,
NEW YORK AND BOMBAY.

NEWCASTLE-UPON-TYNE: ANDREW REID & COMPANY, LIMITED.

1898.

LONDON AND NEWCASTLE-UPON-TYNE
ANDREW REID & COMPANY, Limited, PRINTERS AND PUBLISHERS

76221

Copyright.

PREFACE.

This Manual has been written to meet a need amongst junior ship-draughtsmen, apprentice loftsmen, and class students. The Author trusts it may answer its purpose, for he has sought to make the subject clear by graphic sketch and everyday shipyard language.

After careful consideration, it was thought it would be almost a repetition to enter very fully into the laying-off of war vessels on the loft floor; attention, however, has been given to explaining some of the more prominent parts which distinguish this class of work.

The Author has to express his obligations to the loftsmen of the yards where he has had the pleasure of being located at different times, whom he has found, without exception, always willing to show the points of their craft. Especial thanks are due to the firm of Messrs. William Doxford & Sons, Sunderland, for early instruction in practical laying-off.

The Author wishes to express his gratitude to the following firms, who have willingly helped in this work by giving access to drawings of special parts:—

    Messrs. SWAN & HUNTER, LIMITED, Wallsend-upon-Tyne.
    Messrs. PALMER'S IRON AND ENGINEERING WORKS, Jarrow-upon-Tyne.
    Messrs. WIGHAM RICHARDSON & Co., Walker-upon-Tyne.
    Colonel H. F. SWAN, J.P., of Messrs. ARMSTRONG, WHITWORTH, & Co., Walker-upon-Tyne.

He is further under obligation to the late Professor Rankine's most excellent and advanced work, *Shipbuilding, Theoretical and Practical.*

Any suggestions tending to make the book more perfect for future editions, will be gladly received by the Author,

                              THOMAS H. WATSON.

NEWCASTLE-UPON-TYNE,
   January, 1898.

CONTENTS.

MERCANTILE VESSELS.

CHAPTER I.—LAYING-OFF THE SHEER DRAUGHT ON THE LOFT FLOOR.

Explanation of the Sheer Draught—Profile, Half Breadth and Body Plans—Lines Composing these Plans—How they appear in the Sheer Draught—True Form of the Lines Particulars given to Lay Ship Down—Midship Section—Explanation of Terms—Length over all—Length between Perpendiculars—Moulded Beam—Moulded Depth—Rise of Bottom—Tumble-home—Deadflat—Turn of Bilge—Camber of Beam—Fore-foot—Depth of Hold—Change of Frames—Classification Length and Depth—Flam—Laying-off the Vessel on the Loft Floor—Mode of Procedure—Drawing in the Profile, Body, and Half Breadth Plans—Fairing-up the Body Plan—The Best Lines for Fairing-up—Projection of Bilge Diagonal—Definition of Fairness—Definition of Fair Line—Use of Diagonals—Raised Keel—Final Test of Fairness—Fairing by Contraction—Definition of a Bow Line and Buttock—Definition of a Bilge Diagonal—Twin Screw Bossing—Twin Screw Bossing, Covered Shaft—Bossed Frames Aft—Finish of the After End—Fairing-up the Stern—Diagonal Line Cutting Knuckle—To Fair Frames from the Transom by a Diagonal—To Fair Frames from the Knuckle by a Diagonal—Sheer Line—Approximate Sheer—Construction of a Sheer Diagram—Another Method of Constructing Sheer Diagram—Lowest Point of Sheer—Ready Method of Finding Sheer Line—Sheer under Freeboard Tables—Messrs. Swan & Hunter's Method of Laying-off ... ... pp. 1–26

CHAPTER II. CANT FRAMES.

Cant Frames in the Fore Body—Projection into the Sheer—True Form in the Sheer—Projection into the Body—Projection into the Sheer on Diagonal Lines—Expansion of the Moulded Edge in the Sheer on Diagonal Lines—Expansion of the Bevelled Edge—Projection into the Sheer on Bow Lines—Expansion of Moulded Edge on Bow Lines—Stern Cant Frames—Projection of the Moulded Edge into the Sheer—Projection of the Bevelled Edge into the Sheer—True Form of the Moulded Edge in the Sheer—True Form of the Bevelled Edge in the Sheer—Lifting Bevels—Moulds ... ... pp. 27–33

CHAPTER III.—DECKS.

Beam Camber Allowed—Method of Laying it Off—To Draw in the Deck at Side Line.—Expansion of the Deck Surface—Method of Laying-off Tapered Stringer Plate—Deck Plate Edges—Wide Stringer Plates ... ... pp. 34–36

vi   CONTENTS.

CHAPTER IV.—FLOORS AND DOUBLE BOTTOMS.

Turned-up Floors—To Obtain the Form—Diminishing Line—Fairing-up—Extreme End Floors—Expansion—Cellular Double Bottoms—To Obtain and Fair the Double Bottom—Expansion of the Inner Bottom—Expansion of the Margin Plate—Obtaining Tank Knees—Abaft and Forward of Double Bottom—Expansion of Double Bottom Floors—McIntyre Tank—Swan Conical Tank ... pp. 37—44

CHAPTER V. SHELL PLATING.

Shell Plating—Obtaining Sight Edges—Lining the Model Off—Fairing the Sight Edges on the Model, on the ¾ Lines, on the Loft Floor.—Transferring the Sight Edges to the Scrieve Board—Ordinary Shell Expansion—Area of the Outer Bottom—Ordering Shell Plating—Stern Expansion—Obtaining True Form of Plating—Check on the Expansion—Another Method of Stern Expansion—Tumble-home Stern Expansion ... ... pp. 45—53

CHAPTER VI.—SCRIEVE BOARD.

Scrieve Board—Information placed upon it—Its Purpose—How Prepared—Scrieving in Frames—Decks—Shell Plating Sight Edges—Shell Plating Inner Edges—Ribbands—Keelsons—Floors—Cant Knees—Lifting Beams—Frame Bevels—Applying Bevels—Checking Bevels—Handy Bevelling Machine—Machine Bevelling ... pp. 54—66

CHAPTER VII.—RIBBANDS AND HARPINS.

Ribbands—Form of a Ribband Line—Stem Termination—Stern Termination—Laying them Off and Marking Battens—Deck and Inner Bottom Ribbands—The Common Harpin—Form of Moulded Edge—Form of Bevelled Edge—The Sheer Harpin—Form of Moulded Edge—Form of Bevelled Edge—Bevelling Board—Expansion of Moulded Edge in Sheer—The Stern Harpin—Form of Moulded Edge—Form of Bevelled Edge ... ... pp. 67—73

CHAPTER VIII.—MOULDS.

Principal Moulds and the Order they are sent into the Yard—Stem—Stern Frame—Shaft Struts—Stern Tubes—Flat Plate Keel—Centre Through Plate Keelson—Boat Beams ... ... pp. 74—84

CHAPTER IX.—POOP ROUND AND TURTLE BACK.

Poop Round—Obtaining Lines and Fairing-up—Expansion of Plating—Turtle Back—How to Obtain and Fair-up the Form—Expansion—Plate Edges... ... ... pp. 85—87

CHAPTER X.—EXPANSION OF STRINGER PLATE AND BEAM KNEES.

Expansion of Stringer Plate with no Sheer, and with Sheer—Template—Allowance for Knees in Ordering Tee Beams, and in Bulb Plates pp. 88—90

CHAPTER XI.—IRON AND STEEL MASTS.

How to Obtain the Form—Expansion—Doubling at Deck—Doubling at Heel—Mast Tube Expansion ... ... pp. 91–95

CHAPTER XII.—MISCELLANEOUS.

Rudder Trunk—Obtaining Form—Expansion—Iron Deck House—Form and Expansion—Cargo Hatch Coamings of Ordinary Type—Form and Expansion—Cargo Hatch Coamings with Bell Mouth Bottoms—Form and Expansion—Marking Off the Hawse Pipes—Cutting Holes—Moulds—Shaft Tunnel of a Single Screw—Obtaining Form—Expansion—Calculation for Round—Marking Off the Freeboard in Screw Steamer, in Sailing Ship—Finding Depth Moulded of a Ship when Dry, when Afloat—Clipper Stem—Trail Board—Figure Step—Lacing Piece—Moulds for Carver—Forecastle head—Setting Off Draught Marks on Stem and Stern—Oval Manhole—Methods of Marking Off ... ... pp. 96–113

WAR VESSELS.
CHAPTER XIII.—ARMOUR.

Protective Deck in a Cruiser—To obtain the Form—Fairing-up the Form—Ordinary Expansion—More Correct Method of Expansion—Mode of Plating—Model of Deck—Bevels for Beams—Belt Armour and Deck in a Battle Ship—General Description of the Structure—Correction on Loft Floor for Belt Armour in Fairing-up the Moulded Form—Belt Deck—Its Support—Connection—Butts and Seams—Belt Armour on Box Framing—Armour Shelf—Protective Deck at Ends in a Battle Ship—Finish of Belt Armour at the Ends—Fairing-up the Belt Armour—Expansion—Moulds Required for Ordering Plates—General Description of the Structure of Barbettes or Redoubts—Expansion of the Armour—Ordering the Armour—Moulds Required—Expansion of Inner Thick Plates—Circular Barbette—Revolving Turret in Redoubt ... pp. 114–126

CHAPTER XIV.—SHELL PLATING AND BILGE KEELS.

Outer Bottom Plate Edges of a Battle Ship—Outer Bottom Plate Edges of a Cruiser—Bilge Keels—Where Best Placed—Bossed Frames Forward in Way of the Ram ... ... pp. 127–129

CHAPTER XV.—DOUBLE BOTTOMS.

To Obtain and Fair the Lines of the Inner Bottom of a Cruiser—Expansion of the Inner Bottom—Expansion of the Longitudinals—Expansion of a Longitudinal on Curved Diagonals—Mocking-up System of Expansion—To Obtain and Fair-up the Inner Bottom Lines of a Battle Ship—Expansion of the Inner Bottom—Bevels on Inner Bottom Frames ... ... pp. 130–140

CHAPTER XVI.—GUN GALLERIES OR SPONSONS.

Obtaining and Fairing Lines of Midship Gun Gallery—Expansion—Obtaining and Fairing Lines of Midship Gun Gallery of Conical Type—Expansion of Conical Type—Semi Egg-shaped Forward Gun Embrasure—Another Form of End Gun Gallery—Expansion of End Gallery ... ... pp. 141–146

CHAPTER XVII.—MOULDS.

Principal Moulds, and the Order they are sent into the Yard—Stern Posts—Stems—Stern Tubes—Struts—Beam Camber—Conning Tower—Pilot Bridge—Boat Davits and Chocks... pp. 147–151

CHAPTER XVIII.—DRAUGHT MARKS.

To Obtain and Line-off the Draught Marks on Stem and Stern pp. 152–153

CHAPTER XIX.—COMPOSITE VESSELS.

Sheer Draught—Extreme Form for the Calculation of Displacement—To Find the Heel of Frames by Approximate Method—Exact Method of Finding Heel of Frames—To Find the Middle Line of Rabbet—To End a Level Line in Half Breadth—To Terminate a Frame in Body—To Find Bearding Line Approximately—To Find the Middle Line of Rabbet Accurately—To Find the Bearding Line by Another Method—Form of Rabbet in Main Keel Piece—Working Base Line ... ... pp. 154–159

CHAPTER XX.—SHEATHED VESSELS.

Alteration of Practice—Thickness of Wood Sheathing—Thickness of Shell Plating—Method of Fastening—Solid Stems and Stern Posts—Hollow Section Stems and Stern Posts—Method of Housing, Planking and Shell Plating—Connection of Keel—Finish of Planking and Plating on the Stern—Stern Post of a Cruiser—Stem of a Cruiser—Shaft Brackets or Struts—Method of Attachment at Heel and Foot—Necessary Moulds—Stern Tubes in Twin Screws—How to House Planking and Shell Plating—Bossing of Frames—Necessary Moulds—Stern Posts in Single Screws—System of Terminating Planking, Shell Plating and Keel—Necessary Moulds—Rabbet in Main Keel Piece—Taking-off Planking and Plating... ... ... pp. 160–164

LIST OF ILLUSTRATIONS.

MERCANTILE VESSELS.

Fig.		Page
1	Sheer Draught of a screw steamer	2
2	,, ,, sailing ship	2
3	Laying-Off on the loft floor ...	2
4	Length between perpendiculars	4
5	Section showing terms used	5
6	Twin screw bossing ...	13
7	,, ,,	14
8	Single screw bossing ...	15
9	Finish of the after end ...	16
10	,, ,, ,, ,,	16
11	,, ,, ,, ,,	16
12	Fairing up the stern ...	19
13	Termination of the diagonals ...	21
14	Sheer plan ...	22
15	,, diagram ...	22
16	,, ,,	23
17	,, ,,	24
18	,, ,,	25
19	,, ,,	25
20	Cants in the fore body ...	28
21	,, ,, ,,	29
22	,, ,, ,,	30
23	,, ,, ,,	31
24	,, after end ...	32
25	Camber form ...	34
26	,, ,,	34
27	,, ,,	34
28	,, ,,	34
29	Deck at side on camber curve ...	34
30	Tapered stringer plate ...	36
31	Deck-plating edges ...	36
32	,, ,,	36
33	Wide stringer plates ...	36
34	Curved floors ...	37
35	Extreme end floors sailing ship	39
36	,, ,, screw steamer	40
37	Cellular double bottom ...	42
38	Double bottom margin plate	43
39	,, ,, expansion	42

FIG.		PAGE
40	Shell plating sight edges	45
41	Marking-off model	46
42	Shell expansion	48
43	Area of the outer bottom plating	49
44	Expansion of a shell plate	50
45	Stern expansion	51
46	,, ,,	52
47	,, ,,	53
48	Scrieve board	54
49	Section showing diagonals	56
50	Lifting beams	59
51	,, bevels of frames	60
52	Frame bevels	61
53	Bevel board	62
54	Bevelling machine	63
55	Section showing application of bevels	63
56	Detective bevel machine	64
57	Bevelling machine	65
58	Machine bevelling	66
59	Common harpin	69
60	Harpin bar	69
61	Application of bevel	69
62	Bevelling board	69
63	Sheer harpin	70
64	Bevelling board	71
65	Stern harpin	73
66	Stem foot connection	75
67	,, mould	76
68	,, foot connection	77
69	Connection of keel and stem	78
70	Strut moulds	78
71	Stern tube moulds	79
72	,, ,, ,,	80
73	Keel scrieve board	81
74	,, ,, ,,	81
75	,, mould	82
76	Centre keelson mould	82
77	Boat beam mould	84
78	Poop round	86
79	Expansion of poop round	86
80	Turtle back	87
81	Expansion of stringer plate	89
82	Beam arms	90
83	,, ,,	90
84	,, ,,	90
85	,, ,,	90
86	Masts	91

FIG.		PAGE
87	Mast heel doubling	93
88	,, deck tube	94
89	Rudder trunk	97
90	,, ,, expansion	98
91	,, ,, ,,	99
92	Iron deck house	100
93	Cargo hatch coamings	101
94	Bell mouth cargo hatches	102
95	Hawse pipes	103
96	Shaft tunnel	104
97	Freeboard marks	105
98	,, ,,	105
99	Marking-off freeboard	106
100	Finding depth moulded at ship	107
101	Clipper stem	108
102	Section through cut-water	108
103	Figure step	109
104	,, head block	109
105	Outline ship on blocks	110
106	Section showing sight board	111
107	Oval	112
108	Forming oval	112
109	Machine for forming oval	112

WAR VESSELS.

110	Body plan, showing protective deck	114
111	Sheer, showing protective deck	116
112	Detail, protective deck connection	114
113	Sheer, protective deck	114
114	Expansion protective deck	114
115	Butts of protective deck plating	118
116	Seams ,, ,,	118
117	Finish ,, ,, at stem	114
118	Box framing behind side armour	119
119	Body section (side armour)	123
120	Fore and aft sectional elevation of end armour	120
121	Cross section of end armour	122
122	Plan of finish of ,, ,,	120
123	Longitudinal sectional plan of side armour	123
124	Laying-off side armour	123
125	Expansion ,, ,,	123
126	Finish of side armour at top	119
127	,, ,, ,, bottom	119
128	Barbette, elevation and plan	124
129	Battle ship's ,, ,,	124
130	Expansion of barbette armour	125
130a	Section through thin plating at seams	125

FIG		PAGE
131	Circular barbette	126
132	Revolving turret	126
133	Battle ship's shell plating edges—body plan	127
134	Cruiser's shell plating edges—body plan	128
135	Horizontal section through forward bossed frames ...	128
136	Body sections, showing double bottom ...	130
137	Fairing lines for ,, ,, ...	131
138	Expansion of longitudinal	131
139	Body for expansion of longitudinal on curved diagonals	134
140	Expansion lines of longitudinals on ,, ,,	134
141	Expansion of longitudinals on ,, ,,	134
142	Body sections, showing longitudinal	134
143	Longitudinal mould	135
144	Battle ship's body plan, showing double bottom	136
145	Lines for fairing double bottom, battle ship ...	138
146	Expansion of inner bottom battle ship ...	138
147	Battle ship's midship section	140
148	Midship gun gallery	142
149	,, ,, of conical type ...	143
150	Semi egg-shaped gun embrasure ...	144
151	End gun gallery	145
152	Stern post mould	148
153	Stem mould	148
154	Stern tube mould	148
155	Conning tower mould ...	150
156	Draft marks in war vessels	153

COMPOSITE VESSELS.

157	Finding heel of frames ...	154
158	Middle line of rabbet	156
159	Boundary line, etc.	156
160	Middle line of rabbet	156
161	Form of rabbet in main keel piece	158

SHEATHED VESSELS.

162	Stern post of cruiser	160
163	Stem ,, ...	160
164	Struts ,, ...	161
165	Palm of strut ,, ...	162
166	Stern tube ,, ...	163
167	Stern post ,, ...	164
168	,, ,, ...	164
169	Section through main keel of cruiser ...	163

NAVAL ARCHITECTURE.

MERCANTILE VESSELS.

CHAPTER I.

Explanation of the Sheer Draught: Profile—Half Breadth—Body Plan—Lines Composing these Plans—How they appear in the Sheer Draught—True Form of the Lines—Particulars given to Lay the Ship Down—Explanation of Terms—Length Over All—Length between Perpendiculars—Midship Section—Moulded Beam—Moulded Depth—Rise of Bottom—Tumble-home—Deadflat—Turn of Bilge—Camber of Beam—Fore-foot—Depth of Hold—Change of Frames—Classification Length and Depth—Flam. *Laying-off the Vessel on the Loft Floor:* Mode of Procedure—Drawing in the Profile—Body and Half Breadth Plans. Fairing-up the Body Plan—The Best Lines for Fairing-up—Projection of Bilge Diagonal—Definition of Fairness—Definition of a Fair Line—Use of Diagonals—Raised Keel—Final Test of Fairness—Fairing by Contraction—Definition of a Bow Line and Buttock—Definition of a Bilge Diagonal—Twin Screw Bossing—Twin Screw Bossing-covered Shaft—Bossed Frames Aft—Finish of the After End—Fairing-up the Stern—Diagonal Line Cutting Knuckle—To Fair Frames from the Transom by a Diagonal—To Fair Frames from the Knuckle by a Diagonal—Sheer Line Approximate Sheer—Construction of Sheer Diagram—Another Method of Constructing Sheer Diagram—Lowest Point of the Sheer—Ready Method of finding Sheer Line—Sheer under Freeboard Tables—Messrs. Swan & Hunter's Method of Laying-off.

LAYING-OFF THE SHEER DRAUGHT ON THE LOFT FLOOR.

The term, **Laying-off**, is used to express the method adopted for enlarging and fairing the vessel's outward form to full size on the loft floor, or on paper in the office, on a larger scale than the original design.

The drawing, from which the particulars are taken, for *laying-off* the vessel on the loft floor is **the Sheer Draught**, which is composed of three plans, the Profile or Sheer, the Half Breadth plan, and the Body plan, showing the moulded form of the vessel, that is, the form to the outside of the iron or steel frames. In wood, composite, or sheathed vessels, this plan is made to the outside of the planking for the purpose of calculating the displacement. It may be to any suitable scale: the usual is a quarter of an inch equal to one foot.

The Profile, usually called the Sheer, is a longitudinal elevation with the stem placed to the right hand side. It shows the sheers of the rail, knuckle, and decks; the positions of the frame stations or sections, and the level or water lines; the form of the ship at the centre line, and at fixed longitudinal vertical planes, parallel to the centre, called bow and buttock lines.

The Half Breadth plan, made for the port side only, represents the form of the rail, knuckle, and deck in plan; the longitudinal horizontal shape of equidistant parallel planes from the top of the keel to the assumed load line, or parallel to a fixed load line; also, the position of the frame stations, the buttocks, and bow lines.

The Body plan is the shape of the vessel at transverse vertical planes, at different frame stations in the length, taken square to the keel and the centre line, upon which is also indicated the decks, rail, and knuckle. In the case of yachts, war vessels, and others with a fixed *trim*, the frames are made square to the load line. These sections show the shape of the vessel for one side only. Those which come abaft of the midship section being shown on the left hand side of the centre line, and those forward on the right hand side.

Fig. 1 is the Sheer Draught of a small screw steamer, with a raised keel, where the frames are square to the load line.

Fig. 2 is that of a sailing ship on a level keel. THIS PLAN IS COMPOSED of the following :—Rail, knuckle, deck, and level or water lines, frame stations or sections, buttock and bow lines, diagonals, and the midship form.

They appear in the drawing as follows :—

Lines	In Profile	In Half Breadth	In Body
Rail	Curved	Curved	Curved
Knuckle	,,	,,	,,
Deck	,,	,,	,,
Level or water	Straight	,,	Straight
Buttocks & bow	Curved	Straight	,,
Diagonals	,,	Curved	,,
Frame stations	Straight	Straight	Curved

Their true form is seen as follows :—

In Profile	In Half Breadth	In Body
Bow lines	Water lines	Frame stations or sections
Buttocks	Diagonals	

Rail, knuckle, and deck.

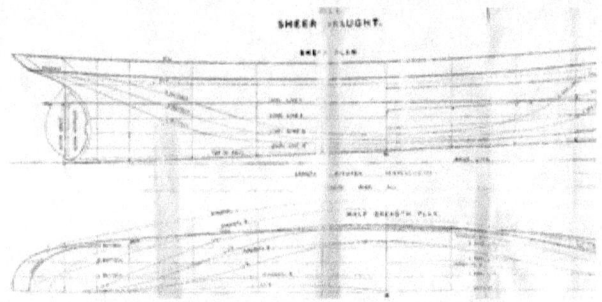

FIG. 2
SHEER DRAUGHT

SHEER PLAN.

HALF BREADTH PLAN.

It is not customary in merchant work to give the loftsman a copy of the Sheer Draught, but the *following particulars are supplied*, in what is called the loft book, to lay the vessel down :—

The length over all, length between the perpendiculars, moulded beam, moulded depth, rise of the bottom, and fall in of the bilge amidships, tumble-home at the deck, size of the forgings (including the keel), particulars of the boss and the centre of the screw shaft, form of the extreme ends in Profile and Half Breadth, sheer heights and half breadths on the rail, knuckle, and deck, half ordinates on the water lines and diagonals, buttock and bow line heights on the ordinate or frame stations, position of the frame stations, water lines, buttocks, bow lines, diagonals, and deadflat, rise of keel (if any), distance apart and size of frame and reverse bars, camber of beam.

Before proceeding further it will be best to give a *brief explanation of the terms used.*

Length over all is the level length from the extreme point of the stern to the fore end of the figure head, or the extreme point of the stem.

Length between Perpendiculars (see **Fig. 4**). 1. Case of perpendicular stem and stern posts. The length is the level distance from the after side of the stern post to the fore side of the stem in line of the upper deck. 2. Case of a raked straight stem or stern post. Produce the fore side of the stem and the after side of the stern post until the lines cut the upper deck. Drop perpendiculars from these points, then the level distance between them is the length. 3. Case of a curved or clipper stem. Produce the lower portion of the stem, below the cut-water, in the same direction until it cuts the line of the upper deck and the same with the after side of the stern post. Drop perpendiculars from these points, then the distance between such is the length. 4. Case of warships and yachts, the length is taken on the load line from the aft side of the stern post to the fore side of the stem."

Midship Section is the fullest part of the ship, and is generally placed midway between the perpendiculars. There are cases where it is nearer the stern than the stem. It is supposed to be at the lowest point of the sheer.

Moulded Beam is the greatest width of the ship from heel of frame to heel of frame on the midship section (see **Fig. 5**).

---

* The length between perpendiculars and depth moulded in an awning-decked ship is invariably taken to the deck below the upper or awning, called the main deck.

It is not customary in merchant work to give the loftsman a copy of the Sheer Draught, but the *following particulars are supplied*, in what is called the loft book, to lay the vessel down :—

The length over all, length between the perpendiculars, moulded beam, moulded depth, rise of the bottom, and fall in of the bilge amidships, tumble-home at the deck, size of the forgings (including the keel), particulars of the boss and the centre of the screw shaft, form of the extreme ends in Profile and Half Breadth, sheer heights and half breadths on the rail, knuckle, and deck, half ordinates on the water lines and diagonals, buttock and bow line heights on the ordinate or frame stations, position of the frame stations, water lines, buttocks, bow lines, diagonals, and deadflat, rise of keel (if any), distance apart and size of frame and reverse bars, camber of beam.

Before proceeding further it will be best to give a *brief explanation of the terms used.*

Length over all is the level length from the extreme point of the stern to the fore end of the figure head, or the extreme point of the stem.

Length between Perpendiculars (see **Fig. 4**). 1. Case of perpendicular stem and stern posts. The length is the level distance from the after side of the stern post to the fore side of the stem in line of the upper deck. 2. Case of a raked straight stem or stern post. Produce the fore side of the stem and the after side of the stern post until the lines cut the upper deck. Drop perpendiculars from these points, then the level distance between them is the length. 3. Case of a curved or clipper stem. Produce the lower portion of the stem, below the cut-water, in the same direction until it cuts the line of the upper deck and the same with the after side of the stern post. Drop perpendiculars from these points, then the distance between such is the length. 4. Case of warships and yachts, the length is taken on the load line from the aft side of the stern post to the fore side of the stem.*

Midship Section is the fullest part of the ship, and is generally placed midway between the perpendiculars. There are cases where it is nearer the stern than the stem. It is supposed to be at the lowest point of the sheer.

Moulded Beam is the greatest width of the ship from heel of frame to heel of frame on the midship section (see **Fig. 5**).

---

* The length between perpendiculars and depth moulded in an awning-decked ship is invariably taken to the deck below the upper or awning, called the main deck.

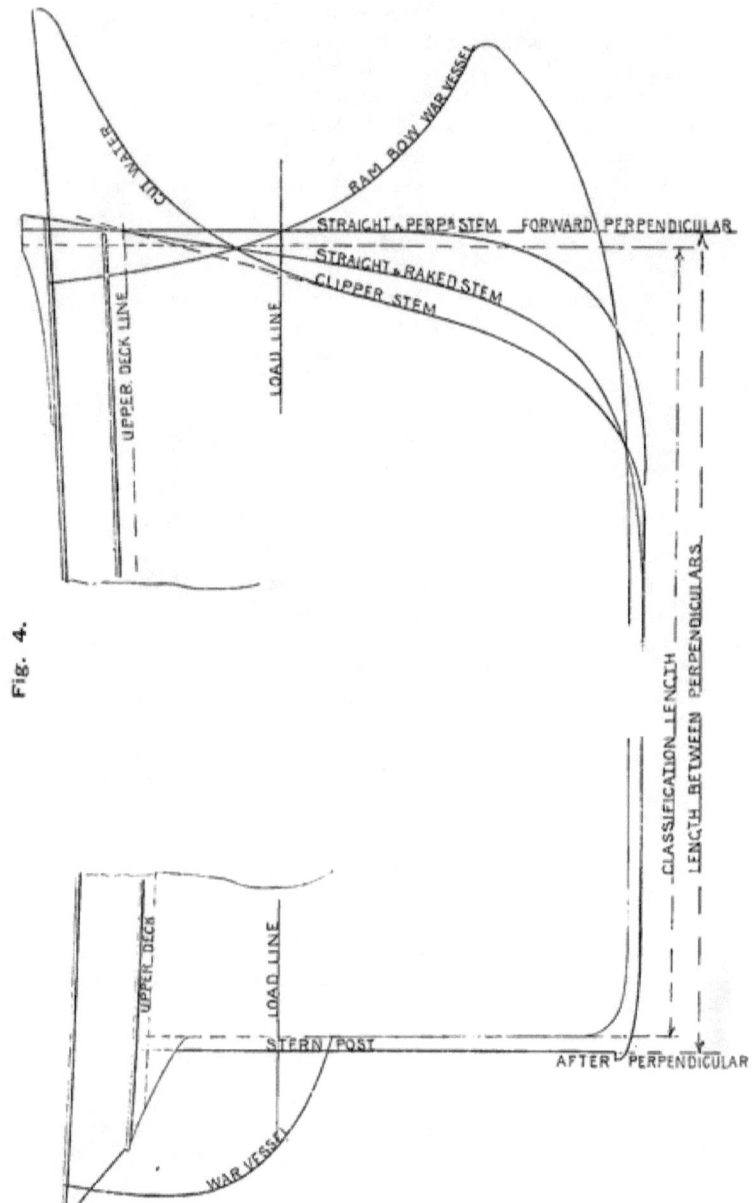

Fig. 4.

Moulded Depth is the vertical distance from the top of the keel—squared out to the side—and the underside of the upper deck stringer plate at the lowest point of the sheer (see **Fig. 5**).

Rise of Bottom.—Produce the line of the midship bottom to the Half Breadth perpendicular, then the distance between this point and the base line—squared out from the top of the keel is the rise of the bottom (see **Fig. 5**). It is, in most cases, straight, but in war vessels or yachts it may be round or hollow.

Fig. 5.

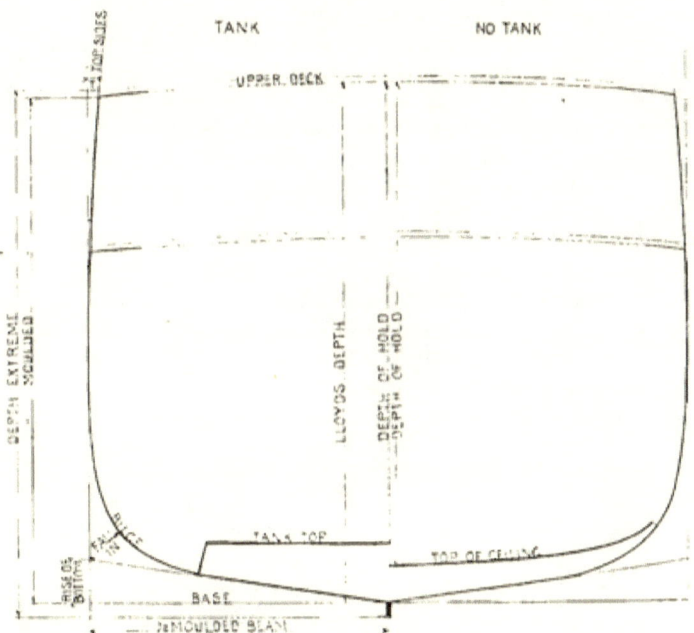

Tumble-home is the distance that the midship side falls in from the half beam perpendiculars in way of the deck (see **T** in **Fig. 5**).

Deadflat, sometimes marked ⊗, is the straightest part of the ship's bottom, or the greatest section amidships, and the point where the frames change.

Turn of Bilge is the curved part joining the ship's side with the flat of the bottom.

Camber of Beam is the "round up" of the deck, sometimes called "crop."

Fore-foot, the part connecting the keel to the stem.

Depth of Hold is the distance at the centre line on the midship frame from the top of the double bottom plating to the top side of the upper deck beams, or, in the case of ordinary turned-up floors, from the top of the wood ceiling (see **Fig. 5**).

Change of Frames.—The bosom of the fore body frames look towards the midships, and the after body frames do the same. The change takes place on the midship frame. This is done because of the excessive bevel which would occur at one end if they all looked the same way, preventing the riveting of the shell flange.

Classification Registry Length is taken in way of the upper deck from the inside of the stern and stem posts (**Fig. 4**). Awning deckers have their length on the deck below the awning deck.

Classification Registry Depth is taken from the top of the keel to the top of the midship beam at the centre line, except in spar and awning-decked vessels, when it is taken to the main deck.

Flam is the opposite of tumble-home, that is, the fall out of the ship's side. Seen in the most of vessels at the fore end.

Having given the meaning of terms used in the loft, we shall now proceed to explain the method of

Laying-off the Vessel on the Loft Floor.—The majority of loft floors are too short to lay the ship down in one length; therefore, it is ordinarily done in two or three lengths by a system of overlapping into each length, which secures continuity of the curves, or the midship part may be taken by itself and faired on the contracted principle explained further on on page 10. Where the loft is sufficiently long to take the vessel in two lengths, each part laps into the other 20 or 30 feet to secure fairness at the juncture, as shown in **Fig. 3**.

Fig. 3 is a drawing showing **Fig. 1** laid-off on the loft floor in two lengths. The trace lines give the cutting points or projections.

Mode of Procedure.—Having obtained particulars of the displacement sections or frame stations, with a figured sketch of the stem and stern from the $\frac{1}{4}$ inch scale Sheer Draught, the first step is to see that the base board is straight, which may form the centre line of the vessel in the plan and the bottom of the keel in the Profile, or a parallel line may be drawn to it for the centre and base.

It may be noted here that the working base line in a vessel where the frames are square to the keel is the top of such keel. In vessels with raised or cambered keels and frames square to the load line, such load line is usually the working base, although some use the line shown in **Figs.** 1 and 3.

Drawing in the Profile.—Divide in on the base the position of the frame stations, and set up the perpendiculars giving the length of the vessel; which may be done with the loft trammels. Then strike in, fore and aft, the top of the keel, which is the working base, the position of the level or water lines, the depth moulded line, and the frame or ordinary stations. Run in with French chalk the inside and outside edges of the stem and fore-foot, afterwards the stern and screw aperture with the counter. Above the depth moulded line on each frame station set up the sheer or deck at side, and run fair line through the spots, terminating on the inside of the stem and at the centre line on the stern. This line may need correcting later on; for it will be evident that the deck sheer at the centre line should be a fair line. For this purpose, when the deck line is run in the Half Breadth, a camber mould is made for the midship beam, as explained on page 34, and the camber due to the *full* width of the deck at each station is set up above the side sheer. Should the spots not give a fair line, the side is lowered or elevated to suit.

The method of finding the sheer intermediate spots when only the heights on the perpendiculars are given is explained on page 20.

The rail and knuckle sheers, also poop and forecastle, if any, are next set off and drawn in.

The Body Plan.—Fix a convenient position on the floor for this, as in **Fig. 3**, and strike in the centre line and the half moulded breadth at each side, rise of the bottom, breadth of the keel, half siding of the stem and stern posts, position of the buttocks, bow lines, and diagonals. Lift off from the Profile the sheers for the rail, knuckle, and deck at side; and place level lines in the Body, through these spots, to take the half breadths—those for the fore body on the right hand side of the centre, and those for the after body on the left hand side. Also lift the height of the frame stations on the inside of the fore-foot, and the same on the after posts, and place them in the Body on the half siding of the posts, which gives terminations for the section feet.

The Half Breadth Plan.—Now square down from the Profile into the Half Breadth the terminations of the rail, knuckle and deck on the stern and stem; forward, these points will be the inside of the stem. Strike in the position of the buttocks and bow lines, and set off on these the form of the rail, knuckle, and deck, aft of the transom; and on the forward terminations lay-off the half-siding of the stem; it may or may not vary. Place the half breadths given you for the rail, knuckle, and deck on their respective stations, and run lines to these spots consistent with fairness; some loftsmen only run the rail and the after knuckle to begin. Then lift from the centre line on a batten these faired widths, and transfer them into the Body on their respective sheer heights.

Measure off in the Body the half ordinates of the water lines, diagonals, the buttock and bow line heights supplied you, which will allow trial frames to be drawn in consistent with absolute fairness. Then you may proceed to

Fair-up the Body Plan.—Square down from the Profile into the Half Breadth, where the water lines cut the *inside* of the stem and stern posts, and set off on these the half-siding of the posts, which will be the terminations of the water lines. Strike in a line amidships showing the half beam, which is a fixed distance out for any line.

Before explaining further, we may say *the best lines for fairing the body are:*—Rail, knuckle, bilge diagonal, alternate water lines—embracing the greatest beam water line—and alternate buttocks and bow lines, or lines which are square to the surfaces to be faired, because they give the nearest true form of the vessel, and, therefore, show the true amount added or taken off to make line fair. The buttocks and bow lines for the ends and under the midship bottom; the water lines for the sides amidships; the bilge diagonal is considered very useful for both purposes, especially under the counter; while a diagonal across the shoulder forward is better than a bow line.

Begin by projecting the Bilge Diagonal into the Profile explained on page 9. This diagonal should be so placed in the Body that it takes the transom, or an additional diagonal used doing so. Square down into the Half Breadth its termination on the inside of the posts. Lift from the Body, *on* the run of the line, the distances from the centre to the half-siding and the different frames, and transfer them into the Half Breadth. Fair line up, and make any corrections in the Body. Several diagonals are shown dotted in

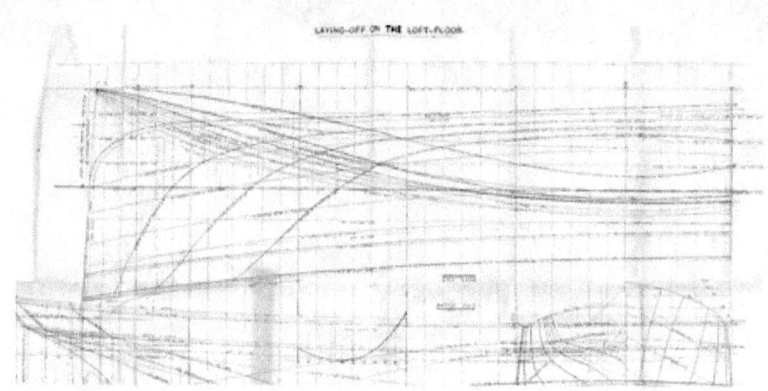

LAYING-OFF ON THE LOFT-FLOOR.

Figs. 1 and **3**, drawn in on this principle. You may now lift half ordinates on the greatest beam water line and place them in the Half Breadth and fair the line up and correct; doing the same with alternate water lines. *When* these lines agree in fairness with the Body sections, lift from the base in the Body the intersection of the frames with alternate buttocks and bow lines, and transfer them into the Profile; and square up from the Half Breadth on to their respective water, diagonal, deck, knuckle, and rail lines, the intersections of the buttocks and bow lines you are dealing with, shown by trace lines in **Fig. 3**, and explained on page 11 under "bow line" Bend a batten round the corresponding spots and fair in. Any alteration to secure fairness must be made in the other plans, and when this is done, and the three plans are absolutely fair on these lines, the remaining water lines, diagonals, buttocks, and bow lines can be run in. If the initial fairing lines are wisely chosen the latter will give little trouble.

Definition of Fairness.—When the buttocks, bow, and sheered lines in the Profile; the level, deck, knuckle, and boundary lines in the Half Breadth; and the frame stations in the Body show continuous curves without abruptness, the ship is fair; or, when spots taken from any two plans and placed in the third show continuity without abruptness. Until this is the case there must be an interchange of spots from one plan to the other to secure agreement and fairness.

Definition of a Fair Line is one pleasing to the eye, or one which has continuity without abruptness; in other words, one which gives a graceful form.

Use of Diagonals.—In many of the large shipbuilding establishments the $\frac{3}{4}$ inch lines are faired principally by diagonals. The first step is to place the diagonals in position in the Body shown in **Fig. 1**, then to project the points a, b, c, d, e, and f, *also* cutting points with buttocks and bow lines into the Profile and draw curves in, from which additional points between the frame stations can be got for the water lines and diagonals in the Half Breadth by squaring down from the Profile where the diagonal or diagonals cut the water lines, and lifting the cutting points, like y^2 in the Half Breadth, from the Body; which is done by measuring the distance *square out* from the centre to where the diagonal cuts the water line y^1, and transferring it on to its respective perpendicular y^2, in the Half Breadth, squared down from the Profile. It will be evident that all the diagonals can be readily drawn thus, in the Half Breadth, in conjunction with the frame

Figs. 1 and **3**, drawn in on this principle. You may now lift half ordinates on the greatest beam water line and place them in the Half Breadth and fair the line up and correct; doing the same with alternate water lines. *When* these lines agree in fairness with the Body sections, lift from the base in the Body the intersection of the frames with alternate buttocks and bow lines, and transfer them into the Profile; and square up from the Half Breadth on to their respective water, diagonal, deck, knuckle, and rail lines, the intersections of the buttocks and bow lines you are dealing with, shown by trace lines in **Fig. 3**, and explained on page 11 under "bow line" Bend a batten round the corresponding spots and fair in. Any alteration to secure fairness must be made in the other plans, and when this is done, and the three plans are absolutely fair on these lines, the remaining water lines, diagonals, buttocks, and bow lines can be run in. If the initial fairing lines are wisely chosen the latter will give little trouble.

Definition of Fairness.—When the buttocks, bow, and sheered lines in the Profile: the level, deck, knuckle, and boundary lines in the Half Breadth: and the frame stations in the Body show continuous curves without abruptness, the ship is fair: or, when spots taken from any two plans and placed in the third show continuity without abruptness. Until this is the case there must be an interchange of spots from one plan to the other to secure agreement and fairness.

Definition of a Fair Line is one pleasing to the eye, or one which has continuity without abruptness: in other words, one which gives a graceful form.

Use of Diagonals.—In many of the large shipbuilding establishments the $\frac{3}{4}$ inch lines are faired principally by diagonals. The first step is to place the diagonals in position in the Body shown in **Fig. 1**, then to project the points a, b, c, d, e, and f, *also* cutting points with buttocks and bow lines into the Profile and draw curves in, from which additional points between the frame stations can be got for the water lines and diagonals in the Half Breadth by squaring down from the Profile where the diagonal or diagonals cut the water lines, and lifting the cutting points, like g^2 in the Half Breadth, from the Body: which is done by measuring the distance *square out* from the centre to where the diagonal cuts the water line g^1, and transferring it on to its respective perpendicular g^2, in the Half Breadth, squared down from the Profile. It will be evident that all the diagonals can be readily drawn thus, in the Half Breadth, in conjunction with the frame cutting points taken from the Body. These diagonals, when so drawn, can also be utilised for the buttocks and bow lines in the Profile; for the cutting points in Half Breadth may be squared up unto the diagonal in the Profile, and also the same points in the Body projected into the Profile. This method will be found of value, when the distance at the ends is great between the load line and the deck.

Fig. 3 shows these diagonals, with level and buttock lines, projected from the Body into the Sheer and Half Breadth. Their true form is also shown. By the trace lines and a pair of compasses the cutting points may be easily found.

Raised Keel.—In this case (**Figs.** 1 and 3) the frame stations are square to the load line, and the top of the keel is the line for the frame feet. Begin by striking in the load line and the perpendiculars square to it, and set down below the load line on the fore perpendicular the draft forward, and on the after perpendicular the draft aft. Connect the two points by a straight line : above this set a parallel line the depth of the keel. Fix the position of the midship frame 6, and on it, above the top of the keel, set up the depth moulded. Draw a line parallel to the load line through this point, and line off above it the sheer heights. Transfer the height of the frame station feet into the Body on the half-siding of keel, stem, and stern posts. It is considered best to use straight water lines in the Body, and in that case, the after sections will come partly below the top of the keel amidships, owing to the keel drooping : otherwise the manner of laying-off is the same as before described, and shown in **Fig. 3.**

Final Test of Fairness.—An excellent manner of testing the fairness of the finished laying down of the moulded form when done by water lines and buttocks, is to project a few diagonals into the Profile and Half Breadth. The system is explained on page 9, under "use of diagonals," and is seen in **Fig. 3.**

Fairing by Contraction.—It is customary in most of the merchant shipyards to fair the midship portion in this way, which requires less room on the loft floor. Diagonal lines, square to the surface of the frames of the Body, are used for this purpose. It may be done on the buttocks, bilge diagonals, and water lines. The method is to contract the frame spacing to $\frac{1}{3}$, $\frac{1}{4}$, or $\frac{1}{2}$ of its actual distance, and upon perpendiculars from these points are set off the *full* distance of each frame station from the base or the half-moulded beam line of the

Body (see A B, C D, E F, and G H respectively in **Fig. 2**). Then a batten is pinned to each set of spots, and a fair curve drawn in. Any necessary correction for fairness is transferred into the Body. This process is correct in principle and greatly facilitates fairing, for the ordinate spots are brought closer together, which increases the curvature: therefore, the batten is more likely to spring fair, owing to its own rigidity, than when the spacing is wide. Care should be taken in fairing the ends to lap them into this portion to give a continuous fair surface. Some establishments fair the midship body up on this principle on the "¾ inch lines," and only lay the *ends of the vessel* off on the loft floor.

Definition of a Bow Line.—A bow line shows in the Profile or Sheer, forward of midships, the moulded shape of the ship at a vertical plane parallel to the longitudinal centre line. Consult **Figs. 1, 2**, and **3**, where it is seen in its true form in the Profile. To draw it you project from the Half Breadth the points of intersection of the line with the water, rail, knuckle, and deck lines on to their corresponding lines; and also level over from the Body into the Sheer to their corresponding stations, the cutting points of the line and the frame stations. It will be seen from this that a bow line must cut level and sheered lines in the same place in the Half Breadth and Sheer; also the frames in the same place in the Body and Sheer. A batten placed on the Sheer spots ought to give a fair line if the Half Breadth and Body are fair, if not, spots are interchanged until the three plans show joint fairness. It is customary in many yards to call this line a buttock.

Definition of a Buttock.—A buttock is similar to a bow line, being a continuation of it, applying to that part abaft of the midship frame.

Definition of a Bilge Diagonal.—A bilge diagonal represents the shape of an oblique longitudinal plane, extending from the centre line of the vessel to the turn of the bilge on the midship frame. See diagonals 2 and 7 in **Figs. 1** and **3** in **Fig. 3**. It may be shown in its *true* form in the Profile or Half Breadth. The points of intersection with the half-siding of the stem or stern post are projected from the Body into the Profile on the inside line of the post, and then squared down on to the base. A batten laid on the run of the diagonal in the Body and the distance, from the centre, of the half-siding of post and the frames marked upon it, and transferred into the Half Breadth on corresponding stations, through which the curve is drawn. This is, perhaps, the best line for fairing, because it cuts the surface usually at about right angles.

Twin Screw Bossing.—Take the simplest case, in **Fig. 6**, where the shaft is parallel to the centre of the ship and the keel line. Draw in the Body the position of the shaft centre a, and from it describe a circle of radius $a\ b$, the required diameter for the tube, etc., i.e., $a^1\ b^1$. Place in the Body the diagonal A B square to the frames cutting a, the shaft centre. Lift the position of the unbossed frames and the centre of the shaft, on the direction of this line, and place them in the Half Breadth which will give $A^1\ B^1$. Run in the centre of the shaft, distance B a, and parallel to the centre of the ship, and arrange tube as shown. To save inside bossing the after end should be fixed to allow nut on the end of the tube to clear ship's side. The length is usually settled by the engineers. Set off line $c\ d$ parallel to the shaft centre, showing the amount of clearance required to draw the tube and stuffing box gland out clear of the frames—allow a little —then set out from this line the width of the frame flange on, say, frame 15. Pin a batten to the line $A^1\ B^1$, and let it spring at the after end to points E, D, C, and fair curve in, keeping the bossed part, consistent with a fair and beautiful form, as short as possible. Lift the distances to this curve from the centre line of the ship, and transfer them on to the line B A in the Body; and with the point a as centre, and these distances as radii describe arcs of circles, connecting them, at the top and bottom, with their respective frames by easy radii or curves. Fair these lines up with close level lines in the Half Breadth, or buttocks in the Profile. Sometimes diagonal lines like F G and H K are used. Their form is shown in the plan by $F^1\ G^1$ and $F^2\ K^1$.

In case the shaft is raked to the keel, and also not parallel to the centre line of the vessel, the lines should be shown in the Profile and Half Breadth, and the centre at each frame placed in the Body, when the above method can be proceeded with.

Twin Screw Bossing—Covered-in Shafts.—In some cases the shaft is covered in from the stern tube, shown in **Fig. 6**, to the after struts, with portable built plate tubes, but this does not give satisfaction for general purposes. The best mode seems to be that given in **Fig. 7**, where the frames are bossed to case in the shafts for the full length, and by it easy access is secured from the inside of the ship. The foremost frames are bossed in the usual way, and gradually formed towards the fine ends into loops with about 2 feet neck at A. Where the outstretch is considerable the frames are continuous,

LAYING-OFF THE SHEER DRAUGHT. 13

Fig. 6.

BODY

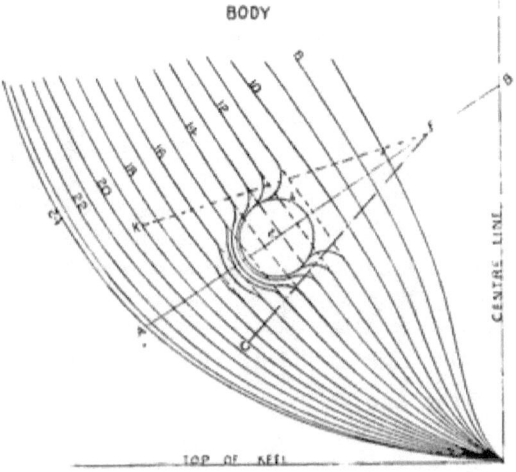

HALF BREADTH

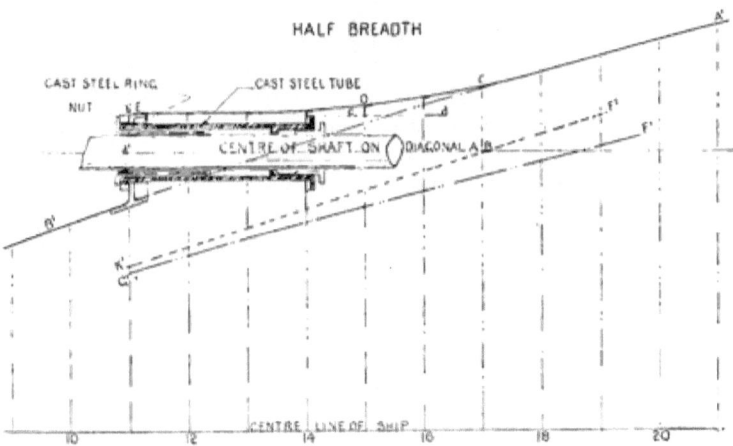

as indicated by dotted lines, and short frames looking aft made to form the loop, which are scarphed at the top and bottom to the ordinary frames. Partial bulkheads or web-plated frames being fitted at intervals to stiffen the structure. It is a common practice to make the after-shaft brackets and the stern post in one piece, and rabbet the shell plating on to it.

To Obtain and Fair the Form in such a case.—The ordinary moulded form is first faired-up on the loft floor, or the ⅜ lines. Then the centre of the shaft is shown on the Sheer, Half Breadth, and Body.

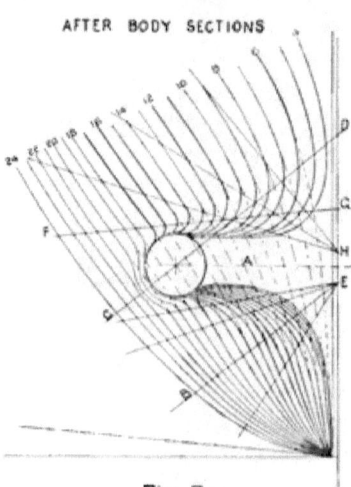

Fig. 7.

A diagonal line C D is placed in the Body and laid off in the Half Breadth, showing the unbossed moulded form of the vessel, and the centre of the shaft as near as possible on this diagonal. The necessary distance outside of the shaft centre, allowing for shafting, framing, and withdrawing of the after tube, etc., is set off from the shaft centre at the different points, and a line run through the spots and faired, which gives the outstretch of the bossing; this is transferred into the Body.

A few approximate sections may now be sketched in, allowing about 2 feet neck at A for getting at the shaft, etc. Diagonal lines G F and E B are put through the curves at the top and bottom, and faired in the Half Breadth. From these lines spots are lifted for the intermediate frames in the Body, and sections drawn. The whole may now be faired by closely spaced radiating diagonal lines from the points H and E, and perhaps a few buttocks. The diagonals should be so placed that they take in the ordinary moulded form with the curves for bossing, to secure a continuous fair surface. These frames are scrieved in on the scrieve board and corrected bevels supplied.

Bossed Frames Aft.—In single screws a few frames are bossed just forward of the screw post. This is ordinarily done in the following

LAYING-OFF THE SHEER DRAUGHT.

manner :—After the frames are scrieved in to the unbossed form, set-off on the centre line of the Body in **Fig. 8** the position of the shaft centre *a*, and from *a* describe semicircle *a c* showing the extreme radius

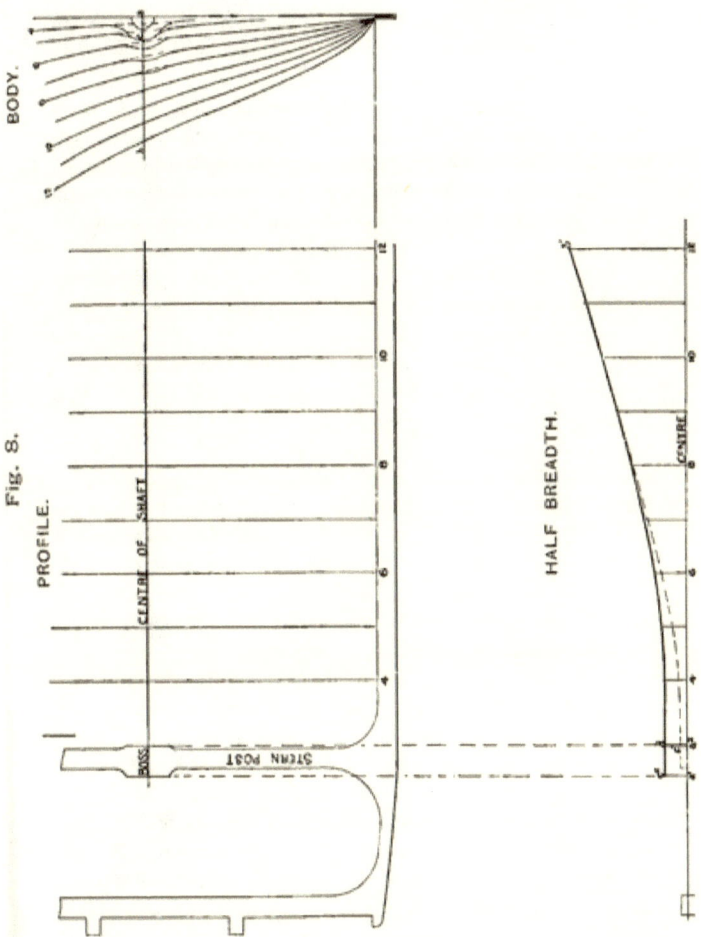

Fig. 8.

of the boss, and draw in the adjacent form, according to the ordered stern frame. Lift the form of the ship to the ordinary curves, shown dotted, on the level line *a b* for about ten frames from the post, and place the distances on their corresponding frames in the Half Breadth, passing curve $c^1 b^1$ through the spots. Draw in this plan the after and forward sides $a^3 c^2$ and $a^2 d$ of the screw post boss, and of the width of radius $a c$. Connect c^2 to d which will be parallel to the centre line. Pin a batten to the original line $b^1 c^1$, and then let the batten spring at the after end sufficient to allow the aftermost frames to clear stern tube, for about two or three spaces, until it joins the point d. Fair the line b^1, d, c^2, and transfer the altered form on to the level line in the Body. Then from the centre of the shaft a, with these points as radii, draw arcs of circles, and connect these arcs with their respective frames at the top and bottom by easy curves. Place level lines through the latter, and test their fairness in the Half Breadth, or fair up with diagonals.

Finish of the After End.—In screw steamers, different forms are adopted on and about the transom foot and the arch of the screw aperture, which occasions a slight difference in the termination of the lines to that of vessels without apertures. In **Fig. 9** is shown one method with a tuck plate flanged on its fore edge to take the shell plating, as seen in the Sheer and section through A B. All the lines terminate on the fore edge of the propeller post produced to meet the counter in a fair curve, and at a distance from the centre equal to the half thickness of the post—L M in the Half Breadth and $L^1 M^1$ in the Body. These points, a, b, c, d, e, and f, are squared down on to L M for the finishing spots for the level lines. The tuck plate is vertical.

In **Fig. 10** the tuck plate is dispensed with; the shell plating edges are run out to the after post, and to break the knuckle, which would be caused by bringing a curved surface abruptly on to a flat one, a small radius is introduced at the line W A X. The shell plates are set on a half-round moulding bar. To obtain the form through W A X, several planes, like A B, are taken square to the line and produced in the other plans, by which the form of the section is got. Produce points A, c, b, and a in the Sheer to corresponding lines in the Half Breadth, namely, A^2, c, b^1, a^1, and draw line $A^2 B^2$. Then produce indefinitely $a a^2, f f^1, b b^2, d d^1, c c^1, e e^1$ square to A B. Lift the cutting point distances of the several lines a^1, b^1, and c square from the centre in the Half Breadth, and lay them out on their corresponding lines in the Sheer, a^1 on $a a^2, b^1$ on $b b^2$, and so on, which will give spots for the section A B. Those for the buttocks may be got by placing the buttock lines parallel to A B, so that they cut their respective trace lines f, d, and e produced. Show in the Sheer

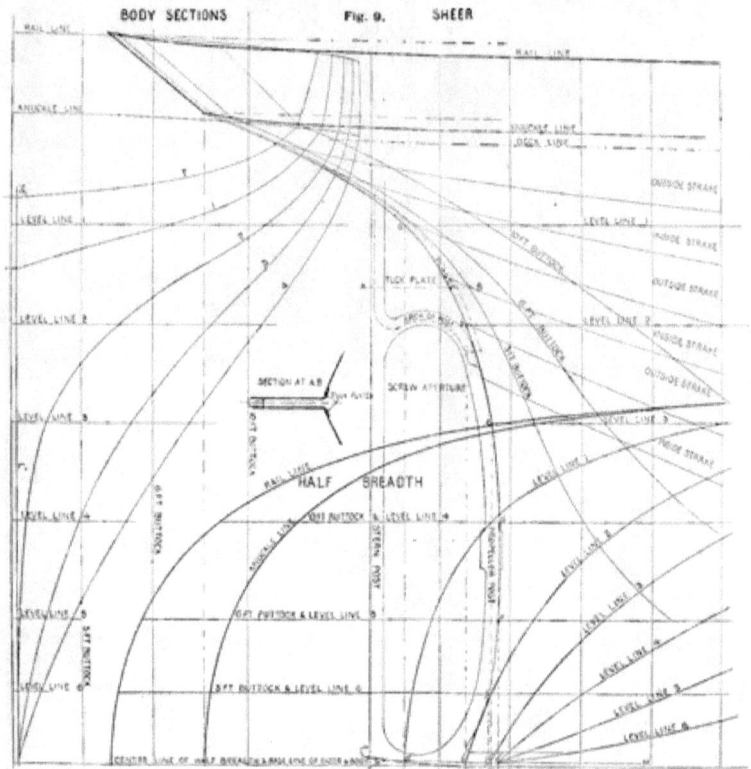

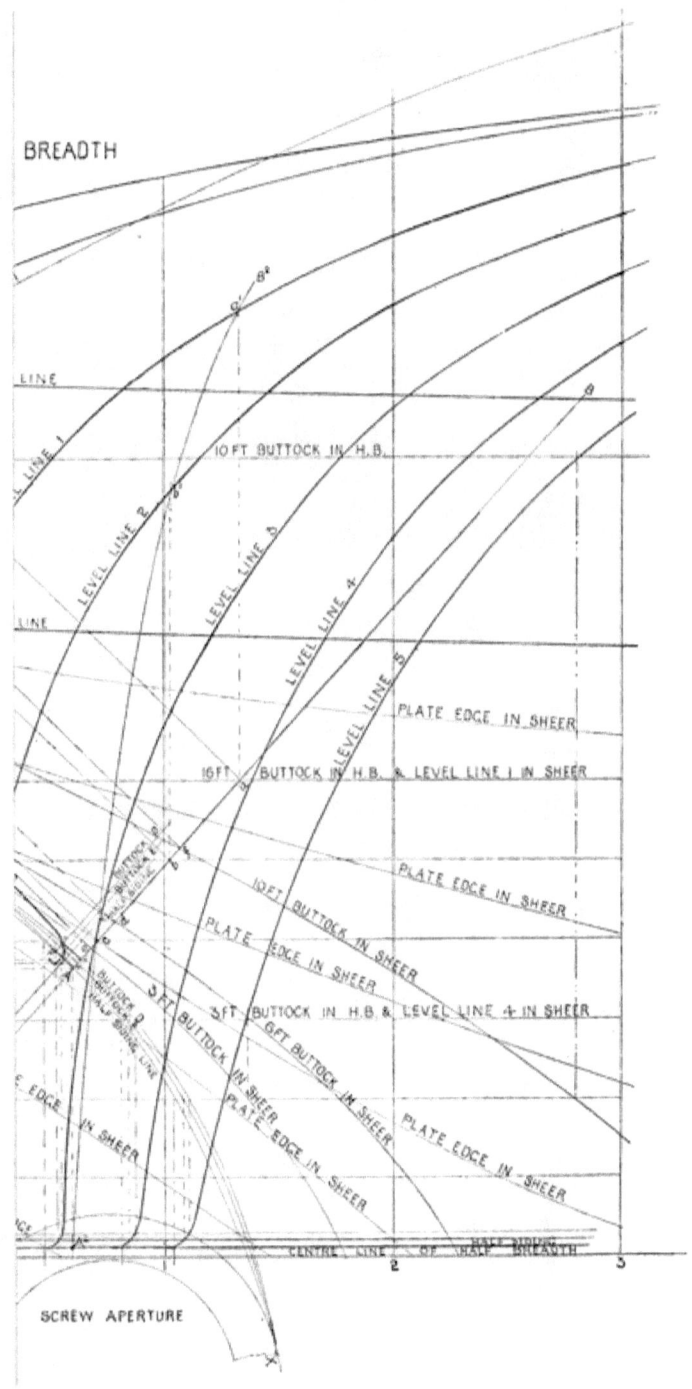

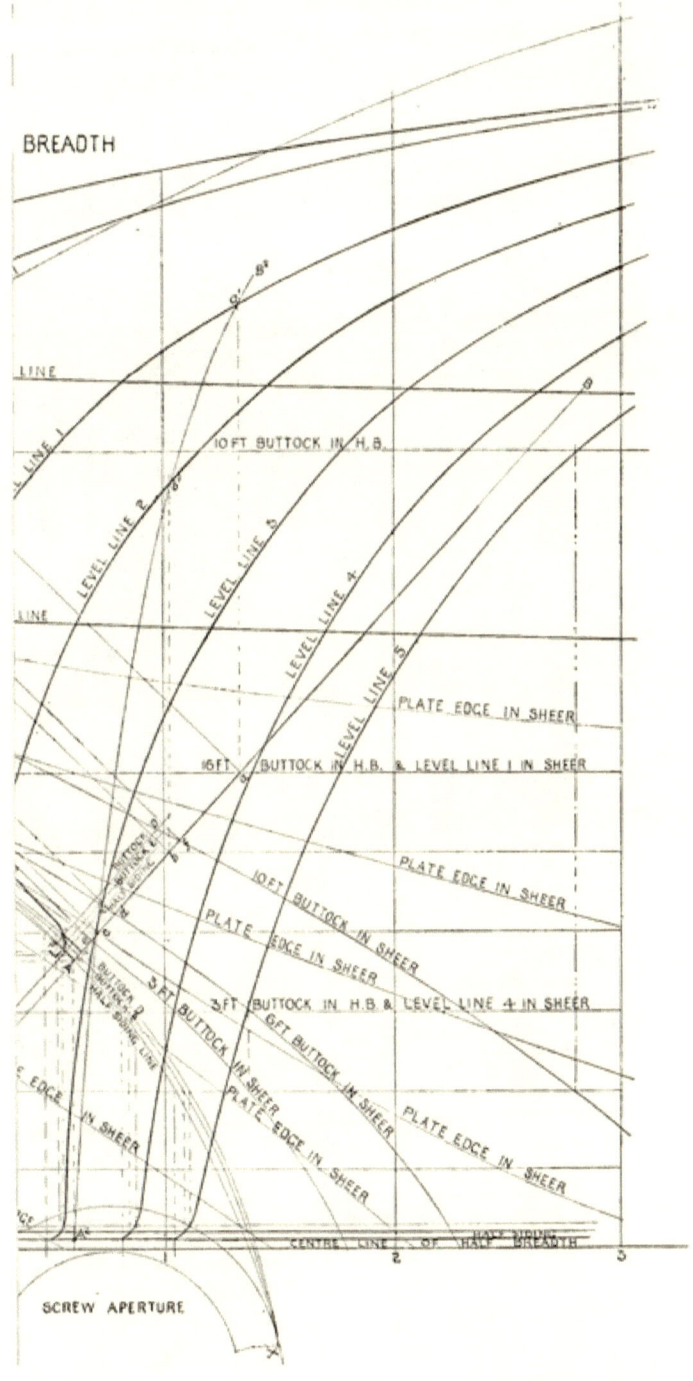

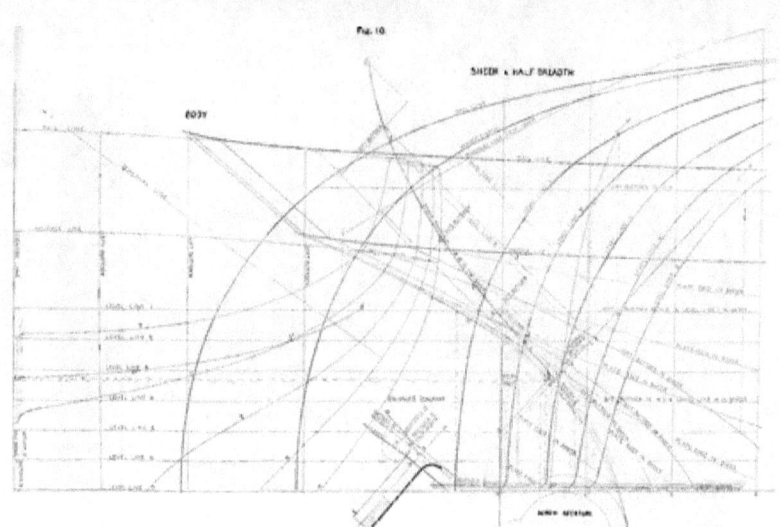

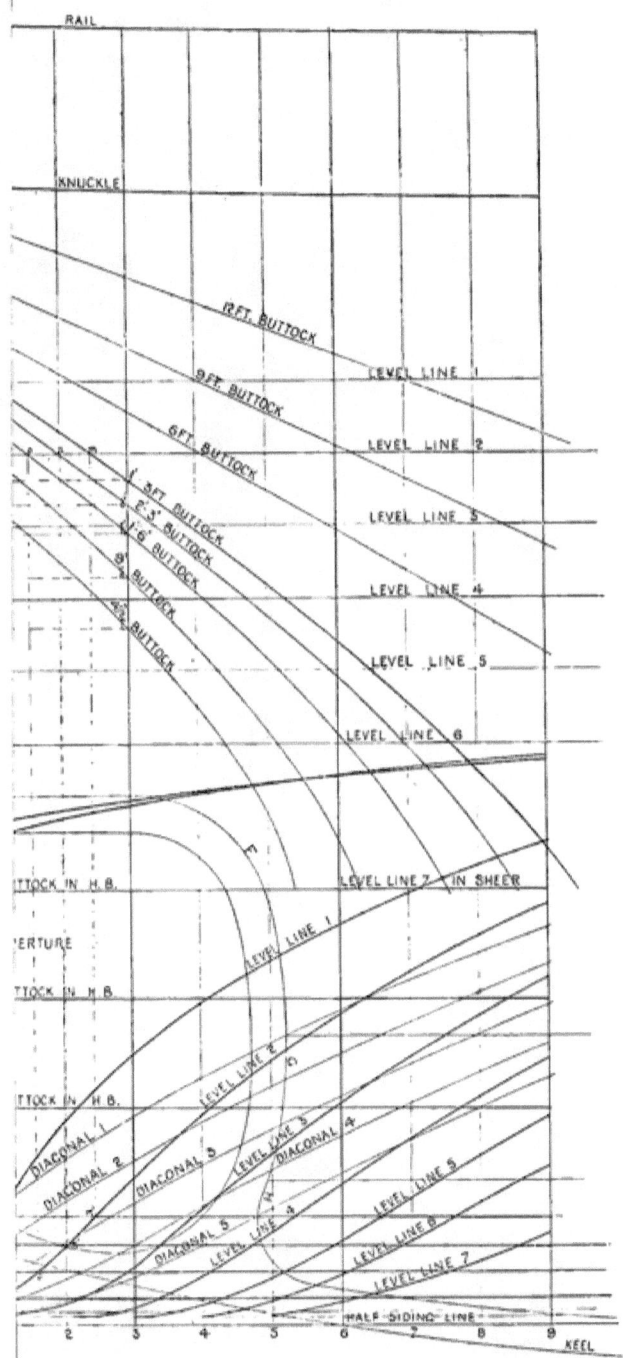

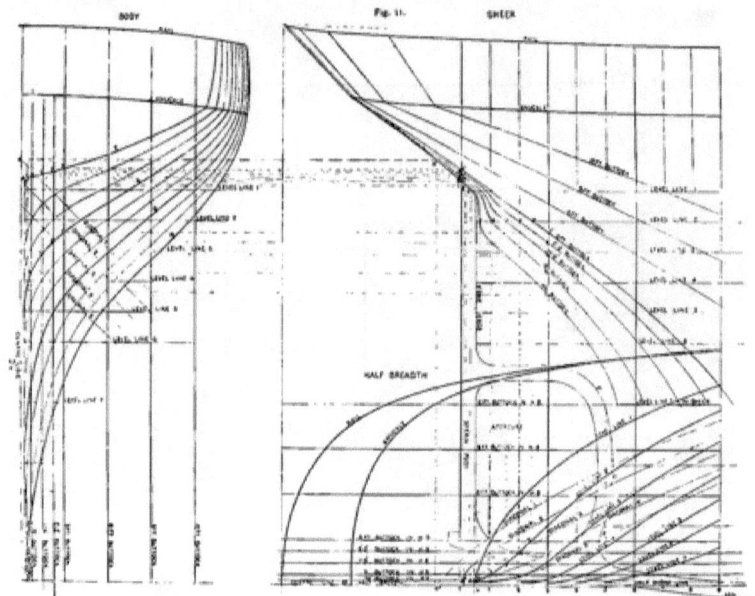

Fig. 11.

the half-siding of the post S L. (See enlarged diagram.) Produce A to cut L, and extend the line indefinitely. With a small radius of 2 to 4 inches, equal P L, describe circle L m n from P to cut L. Then the line S L R is joined with a fair curve to e^1, c^1, d^1, b^2, f^1, etc., which gives form of plane A B. Referring to enlarged sketch, place close buttock lines in the section, D and E, and produce their cutting points with S L R square on to A B, m to m^1 and n to n^1; then m^1 and n^1 are spots for the close buttocks on A B. Buttocks lines may be lined in above and below this section A B, cutting m^1 and n^1, and gradually reducing the radius: or it may be taken the full depth from the transom to the arch of the post. It will be evident that if the close buttocks be placed in the Half Breadth the form of the level lines on these buttocks may be found by squaring down the cutting points, as shown by trace lines. The same applies to the frame stations: the intersections are levelled over on to corresponding buttocks in the Body, and the form of the section drawn in.

In many twin screws the propeller blades cross the centre line of the vessel, at a short distance from each other, which necessitates a shallow aperture, shown in **Fig. 11**. The plate lines are carried to the stern post immediately above the arch of the aperture; and the level lines terminate on the half siding of the fore edge of the stern post, except in such a case as level line 1 which terminates at e.

Perhaps the easiest way to get the suitable form of the level lines is to place in the Body a few closely spaced diagonals, marked 1, 2, 3, 4, 5, square to the curve of the frames, about the run of the plate edges, which readily catch the eye in looking along a ship's after form. Lift the *true form* of these diagonals into the Half Breadth (as explained on page 11), and fair them carefully. The terminating points may be got by following the trace lines, a to a^1, a^1 to a^2, d to d^1, d^1 to d^2 on the half siding of post: and e on the half siding for 3, 4, and 5. When these lines are faired-up, the form of every section to about six frame is transferred into the Body on these diagonals, and curves drawn through the spots. Close buttock lines are then placed next the centre in the Body and Half Breadth, and the cutting points, like A, B, C, D, E on T, and l, k, h, g, f on 3 frame, levelled over into the Sheer on their corresponding frame stations: which, with the terminations on the knuckle squared up from the Half Breadth, enable the form of the buttocks to be drawn in. The fairness may be tested by level lines in the Half Breadth, squaring down the points as in the case of level line 2, r, p, o, n, and m, giving

the half-siding of the post S L. (See enlarged diagram.) Produce A to cut L, and extend the line indefinitely. With a small radius of 2 to 4 inches, equal P L, describe circle L m n from P to cut L. Then the line S L R is joined with a fair curve to e^1, c^1, d^1, b^2, f^1, etc., which gives form of plane A B. Referring to enlarged sketch, place close buttock lines in the section, D and E, and produce their cutting points with S L R square on to A B, m to m^1 and n to n^1: then m^1 and n^1 are spots for the close buttocks on A B. Buttocks lines may be lined in above and below this section A B, cutting m^1 and n^1, and gradually reducing the radius; or it may be taken the full depth from the transom to the arch of the post. It will be evident that if the close buttocks be placed in the Half Breadth the form of the level lines on these buttocks may be found by squaring down the cutting points, as shown by trace lines. The same applies to the frame stations: the intersections are levelled over on to corresponding buttocks in the Body, and the form of the section drawn in.

In many twin screws the propeller blades cross the centre line of the vessel, at a short distance from each other, which necessitates a shallow aperture, shown in **Fig. 11**. The plate lines are carried to the stern post immediately above the arch of the aperture; and the level lines terminate on the half siding of the fore edge of the stern post, except in such a case as level line 1 which terminates at e.

Perhaps the easiest way to get the suitable form of the level lines is to place in the Body a few closely spaced diagonals, marked 1, 2, 3, 4, 5, square to the curve of the frames, about the run of the plate edges, which readily catch the eye in looking along a ship's after form. Lift the *true form* of these diagonals into the Half Breadth (as explained on page 11), and fair them carefully. The terminating points may be got by following the trace lines, a to a^1, a^1 to a^2, d to d^1, d^1 to d^2 on the half siding of post: and e on the half siding for 3, 4, and 5. When these lines are faired-up, the form of every section to about six frame is transferred into the Body on these diagonals, and curves drawn through the spots. Close buttock lines are then placed next the centre in the Body and Half Breadth, and the cutting points, like A, B, C, D, E on T, and l, k, h, g, f on 3 frame, levelled over into the Sheer on their corresponding frame stations: which, with the terminations on the knuckle squared up from the Half Breadth, enable the form of the buttocks to be drawn in. The fairness may be tested by level lines in the Half Breadth, squaring down the points as in the case of level line 2, r, p, o, n, and m, giving points r^1, p^1, o^1, n^1, and m^1. The position of the frames on the level lines is also lifted at the same time from the Body, and placed in the Half Breadth. Before drawing in the lines the termination of each level line is squared down from the Sheer on to the half siding, c to c^1 for level line 1, and e the fore edge of the stern post for the remainder above the arch of the aperture. By this means it will be evident that the true form of the level lines may be found. The close buttock lines have a peculiar character in the Sheer, but this will appear the most reasonable form when consideration is given to the fact that you are dealing with a curved surface in conjunction with a flat one.

In some cases the buttock lines in the Sheer have an easier sweep below the transom, regulated somewhat by the sight edges of the plate lines.

Fairing-up the Stern.—The stern overhung is sometimes faired-up, after the main Body is finished, on the buttocks and a few temporary cross sections taken abaft of the transom. It has to be done very carefully, for a badly faired stern spoils the look of any vessel, or at the least occasions unnecessary packing. **Fig. 12** shows part of the Sheer Draught on the loft floor. Square up from the Half Breadth into the Sheer the buttock intersections A, B, C^1, and D on the rail and knuckle, then run the centre lines in the Sheer E F and G H. The curves at the sides E J and G K are got by finding the amount of camber from the beam mould in way of each buttock and setting it down on the buttock perpendicular from the rail and knuckle centre lines. This camber may easily be found by setting out on the edge of the mould at each side of the centre the square distance of the buttock, and stretching a line from the two points, then the camber is the distance on the centre from this line to the top of the mould. For instance, the camber on J F is $J^1 F^1$. Draw the buttock lines in straight from the rail to the knuckle, and fair them in the Half Breadth by placing level line L M in the Sheer and Body plans, and projecting into the Half Breadth the points a, b, c. Then lift on the level line $L^1 M^1$, from the centre, the distances of O, T, 1 and 2 frames, and place them in the Half Breadth. All these spots are usually adhered to, except that on the buttock next the transom which is made slightly fuller to form a fair line with the other part of the vessel. When faired the corrected points are projected into

NOTE.—In future reference the word Profile will be dropped, and "Sheer" used when referring to it, which is the common term.

the Sheer, and the alteration to buttocks made. The side lines for rail and knuckle abaft of T are placed in the Body by lifting on the buttocks and stern the vertical distances of the rail and

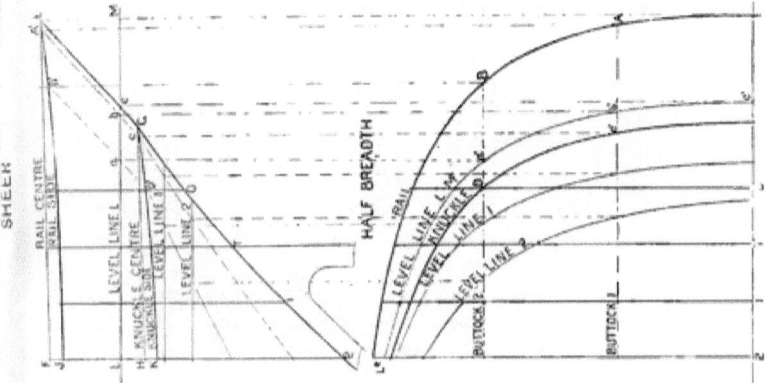

knuckle side sheers, above and below the level line L M, and placing them on their corresponding lines in the Body, through which curves are drawn.

The deck line may be put in, in the same manner, by drawing in the centre, and finding the camber in the usual way. The part below the knuckle line is faired by a few temporary square frames placed abaft of the transom. The heights of the same on the centre, buttocks, and knuckle transferred into the Body, together with the half breadths on the knuckle and rail from the plan, will enable the sections to be drawn in approximately. They are then transferred, on a few closely spaced level lines marked 1 and 2, into the Half Breadth; and faired-up on the buttocks in the Sheer in the usual manner. Care must be exercised to avoid any knuckle on the centre line of the stern; and also to lap the buttocks fair into the part forward of transom already faired-up, which may necessitate a slight alteration in the form of the transom. This method is carried out in some yards before the fairing up of the main body, as part of the first steps.

Diagonal Line Cutting After Knuckle.—To project the diagonal C A, shown in **Fig. 13**, from the Body into the Sheer and Half Breadth. Lift the distances square out from the centre line to the cutting points on the frames, and lay them off on their respective stations in the Half Breadth, D B on $D^1 B^1$, B^1 is the termination on the knuckle line. A curve drawn through spots should be fair, if the sections are fair. This line may be projected into the Sheer by squaring up the point B^1 to knuckle B^2, and E^1 on to the buttock E^2, and levelling over from the Body the intersection of the different frames, shown by dotted lines. Through the spots run curve, and you have the projection of the diagonal.

To Fair Frames from the Transom by a Diagonal.—This is sometimes done by placing line e F in the Half Breadth, $e^1 f^1$, parallel to the centre. Then lift from e on the run of the diagonal the position of the frames, and set them out from $e^1 f^1$. The line D shows the true form of the ship from the transom forward, and should be a fair line if the body is fair : but the best plan is

To Fair the Frames from the Knuckle by a Diagonal.—Lift D B in the Body square out, and set it out on the knuckle $D^1 B^1$, and through the point B^1 draw perpendicular $D^1 B^2$. Then lay a batten on C A, and mark distances C to B, C T, C 3, etc., and lay them out in the Half Breadth, C B on $D^1 B^2$, which gives B^3 for termination, the others on their respective frames. Draw line through spots, which should be fair if the sections are fair. This is the true form of the vessel on the diagonal. If the line is not fair, fair it, and transfer alterations into the Body, and make correction on the buttocks of the Sheer.

Sheer Line.—The mean sheer of the deck at the side for different classes of vessels is given in the " Freeboard Tables," but an approximate mean sheer in inches may be found by dividing the length in feet between the perpendiculars by 10 and adding 10. Take a ship 200 feet long, then the mean sheer is 30 inches, and the united sheers of the after and forward perpendiculars is twice 30, which gives 60 inches. The sheer aft is usually made one-third of the total which leaves two-thirds for forward. In the example taken it would be 20 inches on the after perpendicular and 40 inches on the forward. The lowest point of the sheer may or may not be on the midship

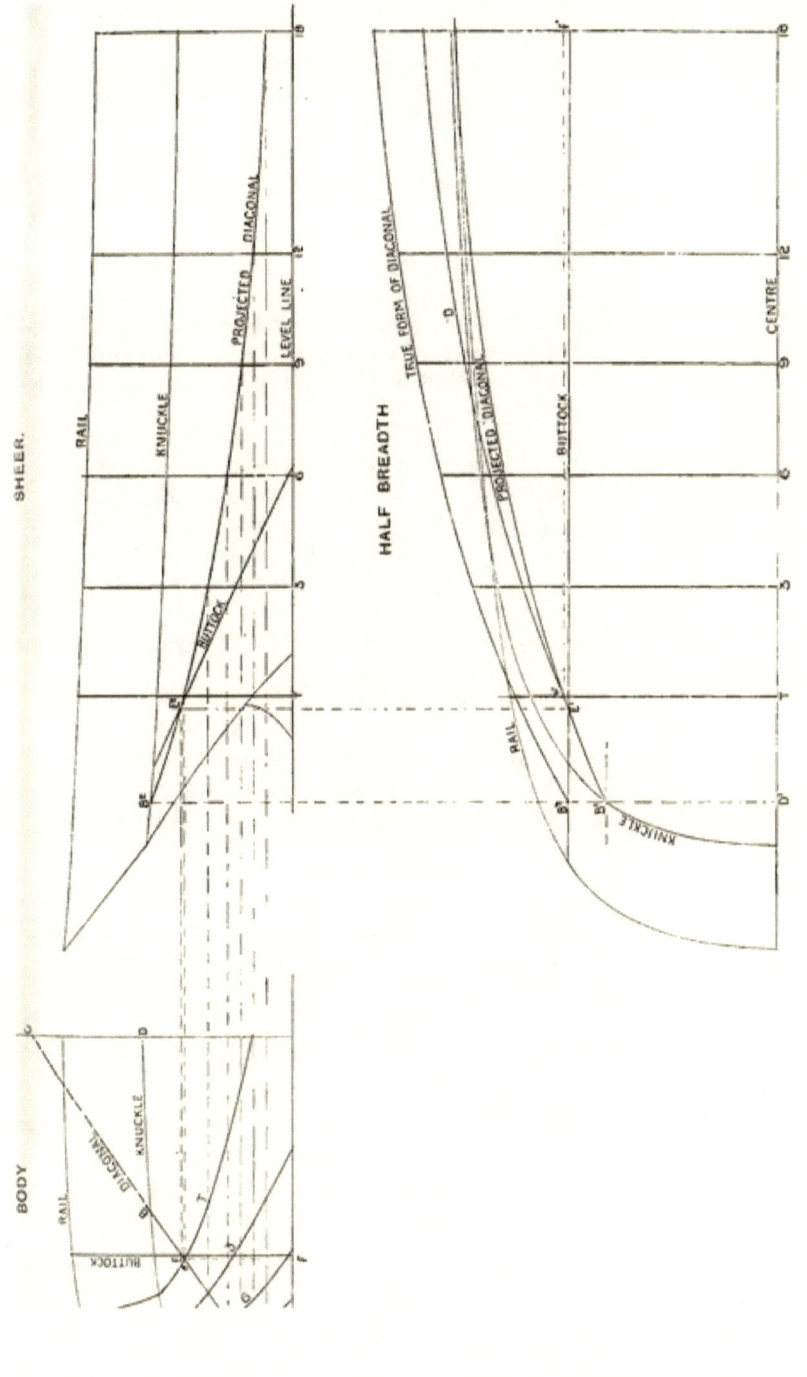

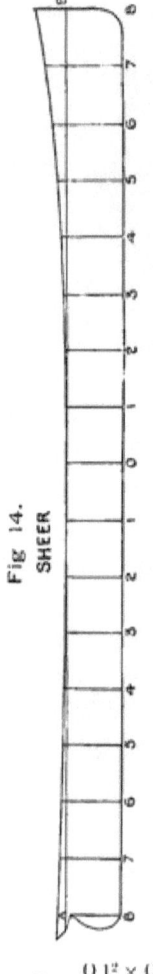

Fig. 14. SHEER

frame. The points in between the lowest point and the perpendiculars are got by two or three different methods, the most useful being a sheer diagram which may be brought into use when once made for any ship, drawn at any scale, within its limits.

Construction of a Sheer Diagram.—Draw in the Sheer plan, **Fig. 14,** the depth moulded line A B extending from the after to the forward perpendiculars, and divide the length into a suitable number of equal spaces, say 16 parts, and produce the points, square to A B, indefinitely above the line. Call the lowest point of the sheer 0, and number the stations abaft and forward of this 1, 2, 3, 4, 5, 6, 7, and 8 respectively. Then in **Fig. 15** on the base C A erect a perpendicular, A B. Let the base C A stand for 0 station, then square each of the numbers 1, 2, 3, 4, 5, 6, 7, and 8, and half each quantity. Treat these quantities as feet, which set up from the base C A on A B to $\frac{1}{8}$ inch scale, and join each point with C. The point C may be found by making the angle E B G equal to 20 degrees, and producing B G to C. This is what is called a sheer diagram. Half sections, if necessary, may be put in for the ends.

Another Way of Constructing a Sheer Diagram.—In **Fig. 15** let O denote the lowest point of the sheer, which is on the depth moulded line C A. Draw perpendicular O P, and make O J equal to O A, and erect perpendiculars J K and A B, equal to the after and forward sheer respectively. Join B to K and produce it to C. Divide O A and O J into, say, 8 equal parts in the points 1, 2, 3, 4, 5, 6, 7. Then the points $a, b, c, d, e, f,$ and g on O P will be found in the following manner:—

$$O\,a = \frac{O\,1^2 \times O\,P}{O\,A^2},\ O\,b = \frac{O\,2^2 \times O\,P}{O\,A^2},\ O\,c = \frac{O\,3^2 \times O\,P}{O\,A^2},\ O\,d = \frac{O\,4^2 \times O\,P}{O\,A^2},$$

$$O\,e = \frac{O\,5^2 \times O\,P}{O\,A^2},\ O\,f = \frac{O\,6^2 \times O\,P}{O\,A^2},\ \text{and}\ O\,g = \frac{O\,7^2 \times O\,P}{O\,A^2}.$$

Join the points $a, b, c, d, e, f,$ and g with C, producing them to A B, and square up on to these lines their corresponding points 1, 2,

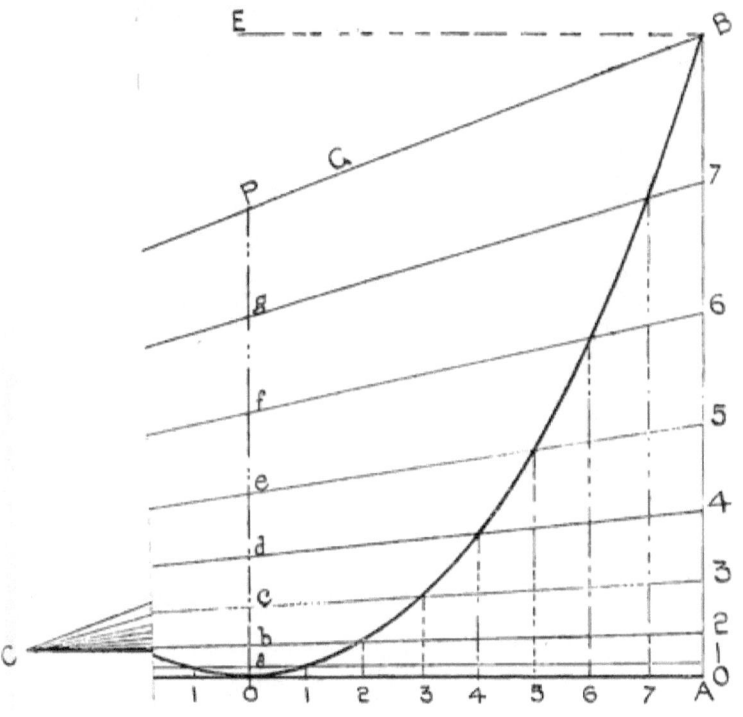

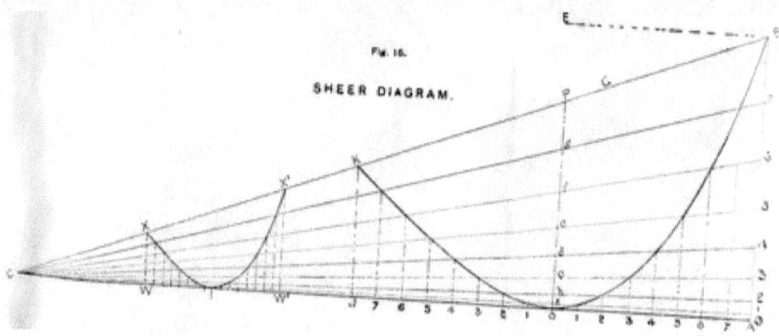

Fig. 16.

SHEER DIAGRAM.

3, etc., as shown, then the line K O B will be the sheer contracted in the length. It can be lifted on to the Sheer in **Fig. 14** above the depth moulded, by taking the distance above C A to each of the points.

It will be readily seen that any sheer, to any scale, within the limits of the diagram can be lifted by finding a point W X, between C A and C B, square to C A, which is equal to the after sheer, and another $W^1 X^1$ equal to the forward. Then divide $W W^1$ into 16 equal parts, and produce points square to C A until they meet their corresponding inclined lines ; then the heights lifted and set off above the depth moulded line will give a perfectly fair line.

Where the Lowest Point is not on the Midship Frame or midway between the Perpendiculars. — This, in some yards, is the case, then the length forward beyond the lowest point will be longer, say, by two spaces than the after distance. Let $a b$ in Fig. 16 equal forward, and $c d$ the after sheer, divide $c a$ into 12 equal parts, let O be the lowest point ; then

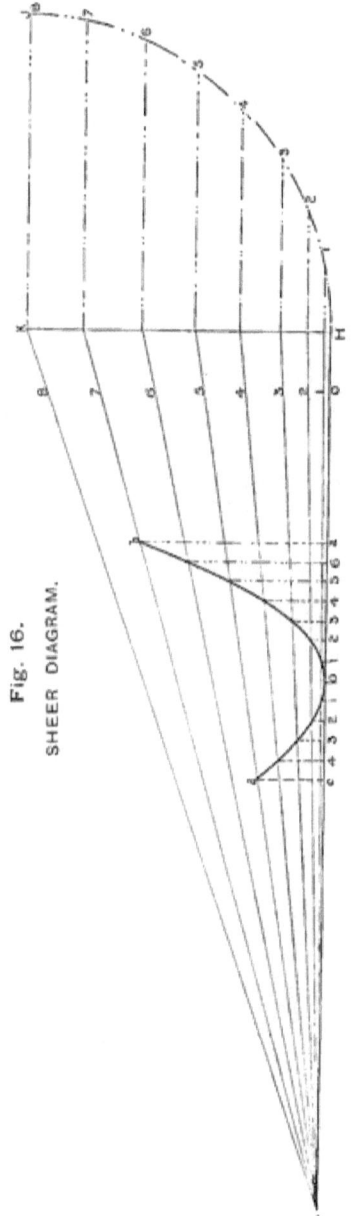

Fig. 16.
SHEER DIAGRAM.

3, etc., as shown, then the line K O B will be the sheer contracted in the length. It can be lifted on to the Sheer in **Fig. 14** above the depth moulded, by taking the distance above C A to each of the points.

It will be readily seen that any sheer, to any scale, within the limits of the diagram can be lifted by finding a point W X, between C A and C B, square to C A, which is equal to the after sheer, and another $W^1 X^1$ equal to the forward. Then divide $W W^1$ into 16 equal parts, and produce points square to C A until they meet their corresponding inclined lines: then the heights lifted and set off above the depth moulded line will give a perfectly fair line.

Where the Lowest Point is not on the Midship Frame or midway between the Perpendiculars. — This, in some yards, is the case, then the length forward beyond the lowest point will be longer, say, by two spaces than the after distance. Let $a b$ in Fig. 16 equal forward, and $c d$ the after sheer, divide $c a$ into 12 equal parts, let O be the lowest point: then

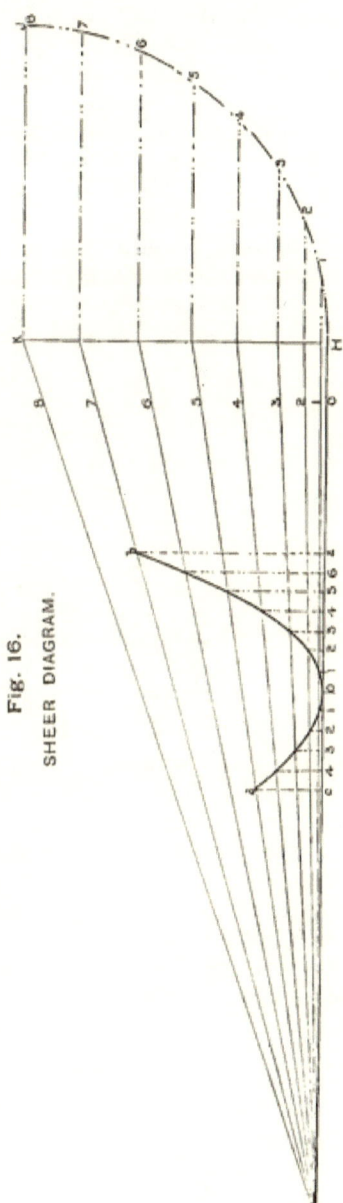

Fig. 16.
SHEER DIAGRAM.

the after length O c will be divided into 5, and O a into 7 equal parts. Produce the points to meet their corresponding lines, the heights of which set off above depth moulded line in the Sheer, which gives you required line.

Another Method of Sheer Diagram is to draw a quadrant of a circle, H J K in **Fig. 16**, with K J equal to the greatest sheer, and divide the arc H J into 8 equal parts, producing the points 1, 2, 3, etc., to H K, as shown. Join each with point P, found in the manner already described. This will give a diagram which can be used in the same manner as the first.

Fig. 17.

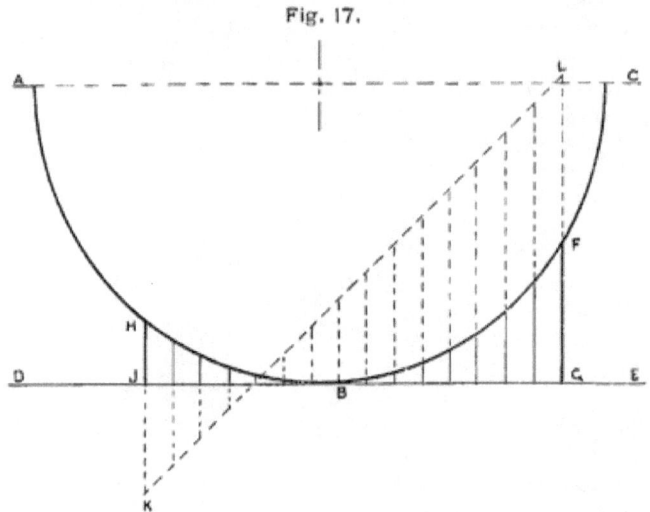

Ready Method of Finding Sheer Points, which is easily done and of considerable use on small scale plans, is to draw a semicircle, A B C in **Fig. 17**, to a suitable scale for the intended plan, and of greater radius than the highest sheer point. Then find a point F G between D E and circle line taken perpendicular to D E and equal to the forward sheer, and one at the other end H J equal to the after sheer. If the distance on D E between J G will divide readily into equal spaces, do so, and erect perpendiculars from the points to the circle line from the line D E, and measure off intermediate sheer points, which set above the depth moulded line in the Sheer plan. If J G does not divide so, then produce G F to L and

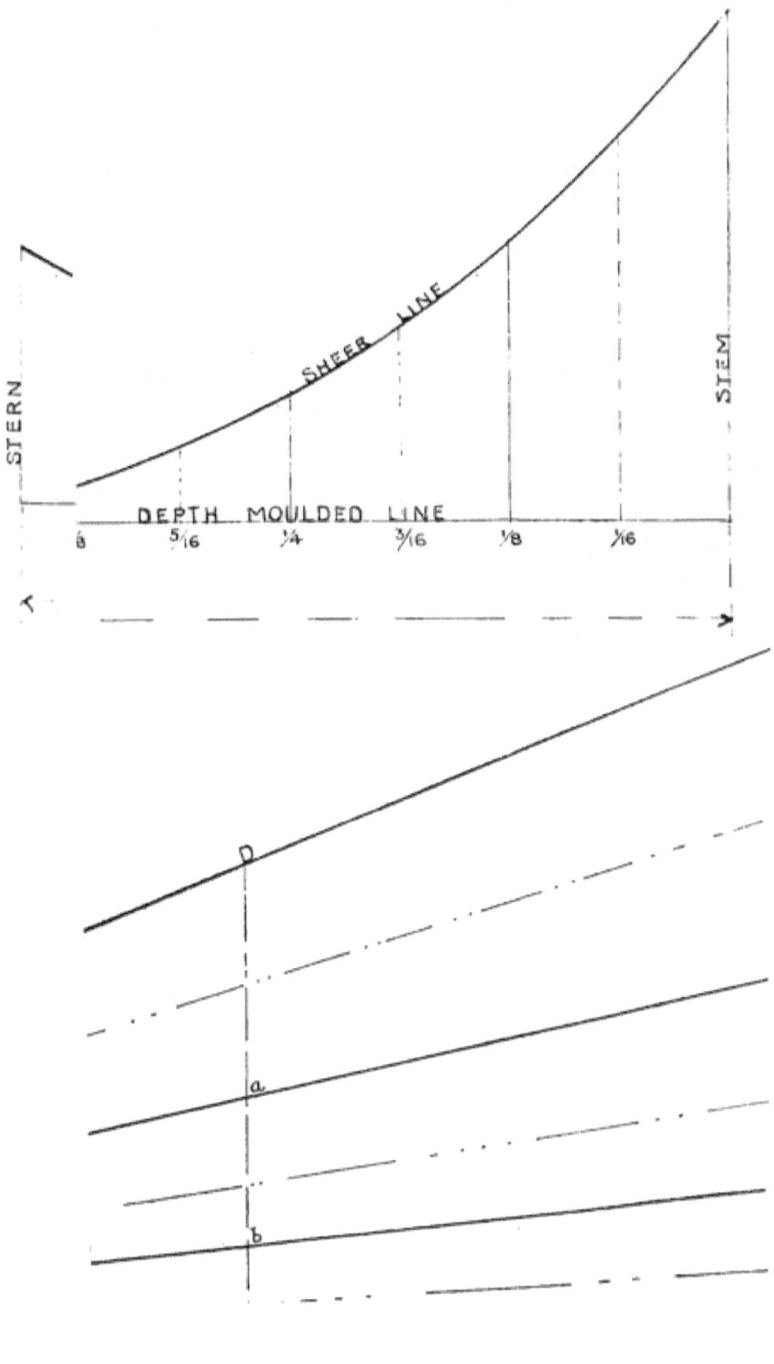

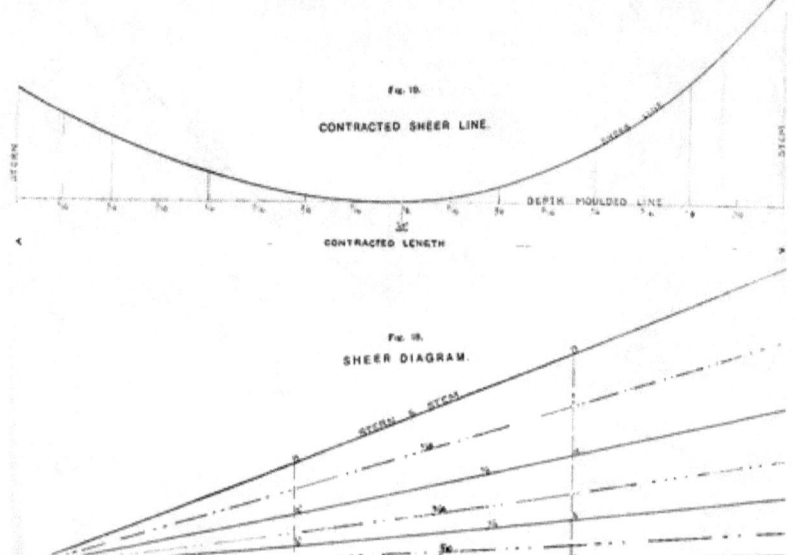

Fig. 19.
CONTRACTED SHEER LINE.

Fig. 19.
SHEER DIAGRAM.

H J to K, and work your scale diagonally until you get a line L K which divides easily, on which mark divisional points and produce them perpendicular to D E as shown. Lift the heights to the arc above D E and transfer to the plan.

Sheer under Freeboard Tables.—In the regulations for freeboard the sheer of the deck at the side is defined as follows :—" At $\frac{1}{2}$ the length of the vessel from the stem or stern post the sheer is to be 55 per cent. of the sheer at stem or stern post : at $\frac{1}{4}$ the length from the stem or stern post 26 per cent., and at $\frac{3}{4}$ the length 7 per cent." The sheer diagram, **Fig. 18**, has been prepared to set off these percentages without the trouble of calculating. As an example, let C D, to scale, equal forward sheer on the stem, and A B after sheer on the stern post. Then the intermediate heights above the depth moulded line in **Fig. 19** are lifted from the base E F in **Fig. 18**; that is, C D set up on the stem, C a at $\frac{1}{8}$, C b at $\frac{1}{4}$, C c at $\frac{3}{8}$, the half length being the touching point. Similarly, A B is the after sheer on the stern post, and the intermediates A a^1 at $\frac{1}{8}$, A b^1 at $\frac{1}{4}$, A c^1 at $\frac{3}{8}$ from the after end. By this means a sufficient number of spots are got for drawing in the line shown contracted in the length in the **Fig. 19**. These divisions in the length are shown further subdivided to $\frac{1}{16}$ of the length, to secure fairness where there is much spread. Any scale may be used on this diagram : and if necessary the radiating lines may be produced to suit 1 inch scale in " laying off " a very long vessel.

The System of Laying-off in vogue at the large establishment of Messrs. C. S. Swan & Hunter, Limited, Wallsend-on-Tyne, is interesting as a novel and successful attempt to overcome some of the practical difficulties involved in laying-off on drawing paper. It is given here because it is an entirely new departure, and as far as the main ideas for its " raison d'être " are concerned it is a decided success.

Some trouble having been experienced in the laying-off department of this and other firms, owing to the gradual shrinkage of the paper causing slight inaccuracies in the subsequent readings taken from it, Mr. G. B. Hunter, chairman of this company, decided to try the experiment of laying-off on some material which should be incapable of shrinkage. The use of wood tables for laying-off was not found to be perfectly satisfactory and accurate, and Mr. Hunter finally decided to try the respective merits of white marble, white cast glass and opal. The principal table for laying-off on 1 inch scale is composed of white marble slabs fitted in lengths of about 6 feet, carefully jointed, with a total length of 30 feet. On this table the Sheer and the

H J to K, and work your scale diagonally until you get a line L K which divides easily, on which mark divisional points and produce them perpendicular to D E as shown. Lift the heights to the arc above D E and transfer to the plan.

Sheer under Freeboard Tables.—In the regulations for freeboard the sheer of the deck at the side is defined as follows:—"At $\frac{1}{2}$ the length of the vessel from the stem or stern post the sheer is to be 55 per cent. of the sheer at stem or stern post; at $\frac{1}{4}$ the length from the stem or stern post 26 per cent., and at $\frac{3}{4}$ the length 7 per cent." The sheer diagram, **Fig. 18**, has been prepared to set off these percentages without the trouble of calculating. As an example, let C D, to scale, equal forward sheer on the stem, and A B after sheer on the stern post. Then the intermediate heights above the depth moulded line in **Fig. 19** are lifted from the base E F in **Fig. 18**; that is, C D set up on the stem, C a at $\frac{1}{2}$, C b at $\frac{1}{4}$, C c at $\frac{3}{4}$, the half length being the touching point. Similarly, A B is the after sheer on the stern post, and the intermediates A a^1 at $\frac{1}{2}$, A b^1 at $\frac{1}{4}$, A c^1 at $\frac{3}{4}$ from the after end. By this means a sufficient number of spots are got for drawing in the line shown contracted in the length in the **Fig. 19**. These divisions in the length are shown further subdivided to $\frac{1}{16}$ of the length, to secure fairness where there is much spread. Any scale may be used on this diagram: and if necessary the radiating lines may be produced to suit 1 inch scale in "laying off" a very long vessel.

The System of Laying-off in vogue at the large establishment of Messrs. C. S. SWAN & HUNTER, LIMITED, WALLSEND-ON-TYNE, is interesting as a novel and successful attempt to overcome some of the practical difficulties involved in laying-off on drawing paper. It is given here because it is an entirely new departure, and as far as the main ideas for its "raison d'être" are concerned it is a decided success.

Some trouble having been experienced in the laying-off department of this and other firms, owing to the gradual shrinkage of the paper causing slight inaccuracies in the subsequent readings taken from it, Mr. G. B. HUNTER, chairman of this company, decided to try the experiment of laying-off on some material which should be incapable of shrinkage. The use of wood tables for laying-off was not found to be perfectly satisfactory and accurate, and Mr. Hunter finally decided to try the respective merits of white marble, white cast glass and opal. The principal table for laying-off on 1 inch scale is composed of white marble slabs fitted in lengths of about 6 feet, carefully jointed, with a total length of 30 feet. On this table the Sheer and the

Half Breadth plans are drawn. The other two tables—one of cast glass and one of opal—are used for the fore and after Body sections respectively. It may be mentioned that the draughtsman in charge of the work prefers the white marble to either of the other tables, as he finds that the glossy surface on the cast glass and opal is against quick working with a drawing pen, and in cold weather the moisture from the hands condenses more rapidly on these than on the marble.

The laying-off is done to a scale of 1 inch to the foot. The system adopted is briefly as follows :—The level lines and frames having been marked off on the Sheer slab, the sheer and half breadth of the rail are faired by contraction and drawn in. They are then transferred to the section slab, thus giving the correct sheer and half breadth at the top of each frame station. The water lines are now roughed in on the Half Breadth and the buttocks faired to them in the Sheer. Then the buttock and water line spots are transferred on to the section table and the cross sections drawn in. Corrections are here made until the buttock, water line, and section spots all agree in absolute fairness. When this stage has been arrived at diagonals are drawn in on the Body section table in convenient positions, as square to the frames as practicable. These diagonals are then lifted off and run on the Sheer slab to further fair-up the sections. The midship portions of the water lines and buttocks are next faired by contraction. The work up to this stage has been done in pencil, but now the corrected lines are inked in. The off-sets are now lifted off and transferred to the loft book for scrieving and future reference. The slabs are easily cleaned with MONKEY BRAND SOAP, and are then ready for fresh work.

CHAPTER II.

Cant Frames in the Fore Body: Projection into the Sheer—True Form in the Sheer—Projection into the Body—Projection into the Sheer on Diagonal Lines—Expansion in the Sheer on Diagonal Lines—Expansion of the Bevelled Edge—Projection into the Sheer on Bow Lines—Expansion of Moulded Edge on Bow Lines. *Stern Cant Frames:* Projection of the Moulded Edge into the Sheer—Projection of the Bevelled Edge into the Sheer—True Form of the Moulded Edge in the Sheer—True Form of the Bevelled Edge—Lifting Bevels—Moulds.

CANT FRAMES.

Cant Frames.—These are always fitted abaft of the transom, and in barges and very full ships are invariably fitted at both ends. By cant frames is meant those which are not square to the centre line of the vessel.

Cant Frames in the Fore Body.—In **Fig. 20** place the heel or moulded edge, A E, of the frame in the Half Breadth as square as possible to the level lines. Set off the bevelled edge G L by girthing from the moulded edge the width of the frame flange on the level lines and rail. Put a curve through the spots : and show on the half siding of the keel M N. It may be noted that the bosom of these frames look towards the centre line of the vessel to give open bevel.

Projection into the Sheer.—Square up the point B on to the top of the keel, C and D to their corresponding level lines, and E to the rail. Do the same with the bevelled edge, H^1 being the termination on the top of the keel squared up from the point H, and L^1 on the rail. Run lines through the spots for each edge, which shows the frame in its *true position.*

True Form in the Sheer.—Draw the level lines $E^1 E^2$ and $L^1 L^2$ through the points E^1 and L^1, the terminations of the moulded and bevelled edges on the rail. Lift the intersections of the moulded edge in the Half Breadth with the half siding of the keel, level lines, and rail, by placing a batten on the run of the line and marking on it the centre A, half siding B, and points C, D, and E. Set these out in the Sheer on their respective lines from the perpendicular A^1 F. That for the rail, A E, set off from F, F E^2, and A B on the top of the keel from A^1 F. Do the same with the bevelled edge, taking care to lift from the line A G, which is square to A E at the centre; G H is the distance for the bevelled edge on the top of the keel in the Sheer, and G L that for the rail on the level line F L². Run lines through the spots, dotting the bevelled edge for distinction. The bevel on the bar at any point is the distance between the expanded edges

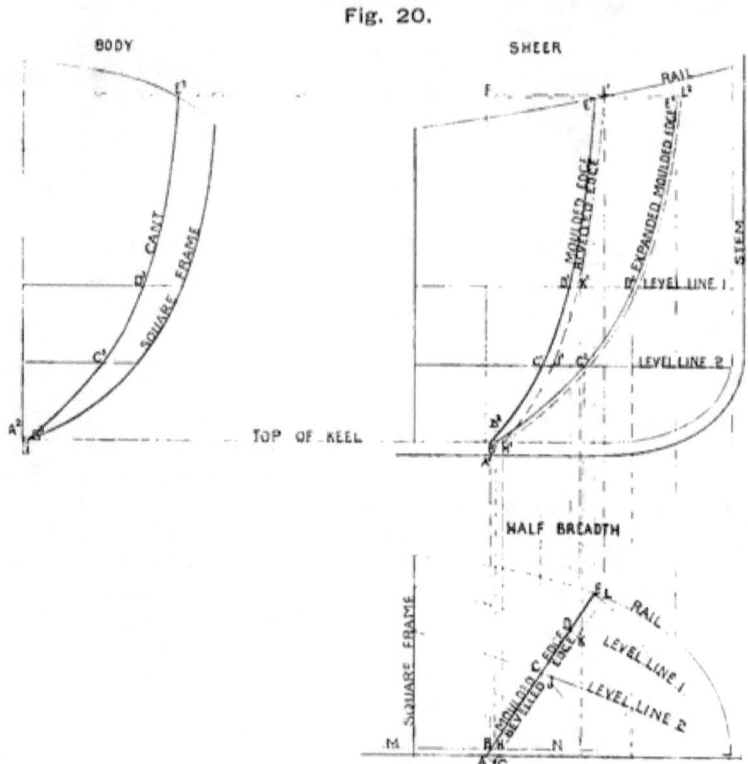

Fig. 20.

taken square to the moulded line. B², C², D², and E² is the true form of the moulded edge, to which the frame must be turned. Wood moulds may be made giving the shape of each frame, with the position of the plate edges, keelsons, decks, and ribbands, accompanied by a bevelling-board showing the bevel at the plate edges: or the true form may be scrieved in on the scrieve-board, with the true position of the ribbands, etc., marked on the scrieves.

Projection into the Body.—Square over into the Body the points B^1 and E^1 for the terminations, and set off on $A^2 B^2$ the half siding, $A^2 B^3$, of the keel at B, then lift off from the Half Breadth, square to the centre line, the intersections of the moulded edge of the cant with the level lines and rail, transferring them on to their respective

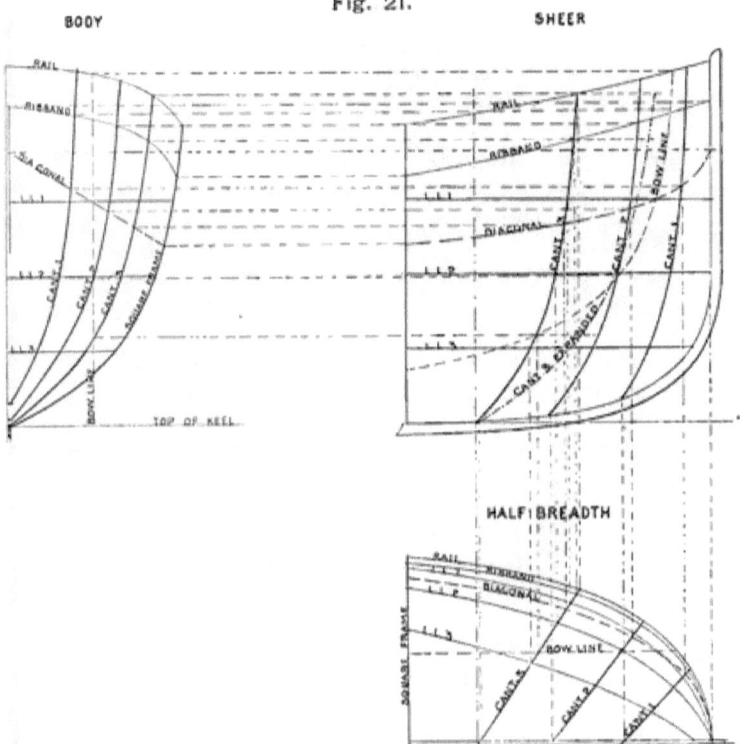

Fig. 21.

level lines in the Body. Pass the curve B^2 and E^2 through the spots which will give the required line. The bevelled edge is projected in the same way; of course, the cutting point on the rail will be slightly higher. The cants may not extend to the rail, but it will be evident that the rail sheer is only chosen to represent any sheered line.

It is sometimes necessary to find the cutting points of diagonals, bow lines, and ribbands on the projected and expanded edges of the cants in the Sheer. **Fig. 21** is, therefore, given, showing in one sketch the diagonals, level lines, ribbands, and bow lines, in conjunction with the forward cants. With a pair of compasses and the trace lines, the method of cutting may be easily understood, but to make the matter perfectly clear it will be explained in detail. The ship is faired-up on square stations before the cants are placed in position. Let it be so in the examples we are considering.

Projection of the Cants into the Sheer on Diagonal Lines.—Show, in **Fig. 22**, the half siding of the keel and stem L M and $L^1 M^1$ in the Half Breadth and Body. Then trace diagonals in the first plan, on the ordinary square frames, by lifting the cutting points square to the centre of the Body and transferring. Their terminations will be got by squaring the intersection points on $L^1 M^1$ over to the inside of the stem in the Sheer, and dropping the points on to the half siding L M of the Half Breadth. Draw the cants 1, 2, and 3 in position in the Half Breadth, after which they are placed in the Body by lifting the cutting points g, h, k, l, m, n, and o on each cant, square out from the centre line, and transferring them in the same manner on to the diagonal lines. The termination at the foot is found by projecting p on to the top of the keel p^2, and from there into the Body on $L^1 M^1$. The other cants are lifted into the Body in the same way. Then square into the Sheer the points p, g, h, k, l, m, n, and o, and produce from the Body the corresponding points p^1, g^1, h^1, k^1, and so on, until they meet, which will give cutting points p^2, g^2, h^2, k^2, etc., for No. 3 cant, and also for projections of the diagonals. If this is done with 2 and 3 cants, the diagonals and cants can be drawn in their true positions by tracing lines through the spots as shown. The bevelled edge may be produced in the same manner.

Expansion of the Cants in the Sheer on Diagonal Lines.—To make it clear only one cant, No. 3, will be taken. In **Fig. 22** produce level lines through the points o^2, n^2, m^2, etc., indefinitely. Draw up perpendicular A B from A. Then lift on the run of the cant the distances A p, A g, A h, A k, A l . . . . . A o, and place them on their corresponding level heights in the Sheer forward of A B, and draw line through the spots, which will give true moulded form of the cant No. 3. A o is set off on B o^2 produced, and A p on the top of the keel produced from p^2.

Expansion of the Bevelled Edge on Sheer Diagonal Lines.—Draw A A^1 square to A O from A. Then lift intersections

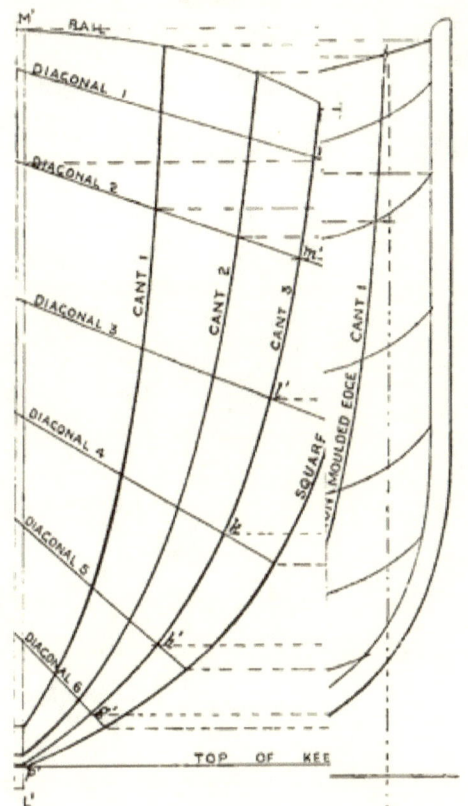
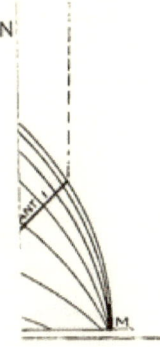

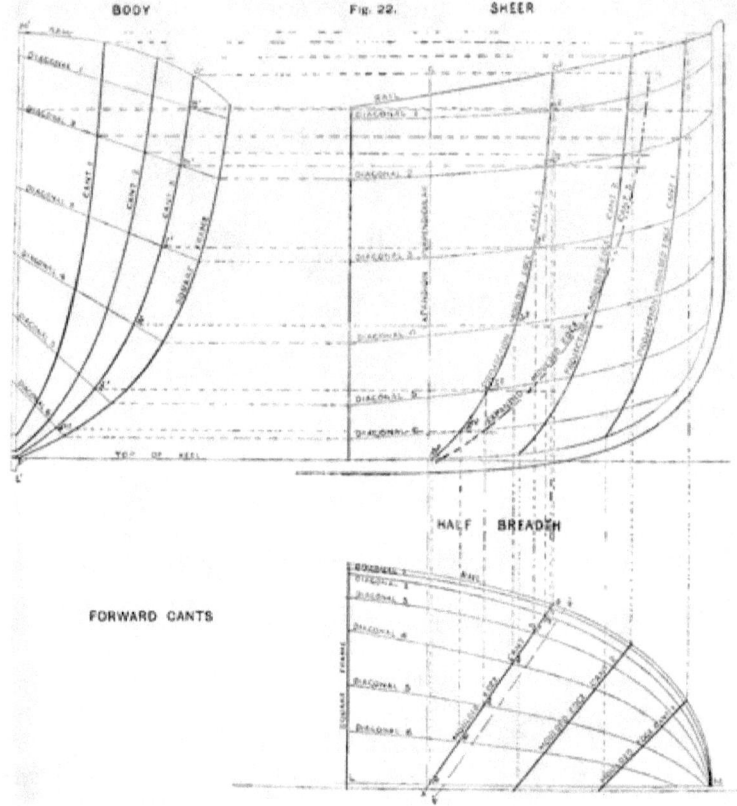

Fig. 22.

FORWARD CANTS

of half siding, etc., from A^1 on the line of the bevelled edge, and transfer into the Sheer on their level heights from A B, passing a curve through the spots, which will give the line of the bevelled edge.

It is usual to place the true form of the cants forward on the scrieve-board with the position of ribbands, decks, plate edges, and

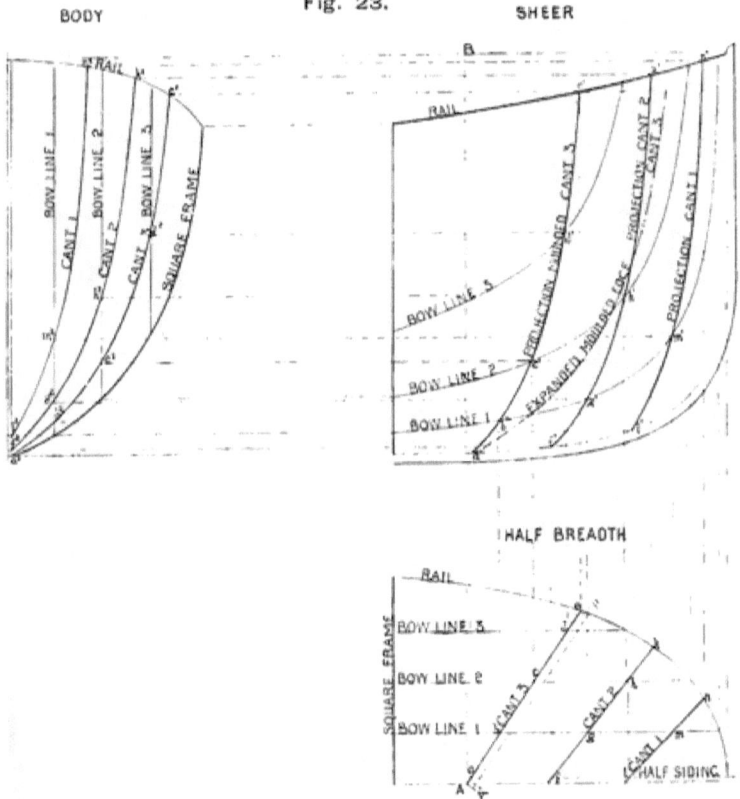

Fig. 23.

keelsons marked upon them. Sometimes a wood mould is made of each with the different markings on and a line for the foot.

Projection of the Cants into the Sheer on the Bow Lines.—Show cants in the Half Breadth and the Body of **Fig. 23** in the manner already described. Square up the points a, b, c, d, and e, until they meet in a^1, b^1, c^1, d^1, and e^1, the corresponding points a^2, b^2,

CANT FRAMES. 31

of half siding, etc., from A^1 on the line of the bevelled edge, and transfer into the Sheer on their level heights from A B, passing a curve through the spots, which will give the line of the bevelled edge.

It is usual to place the true form of the cants forward on the scrieve-board with the position of ribbands, decks, plate edges, and

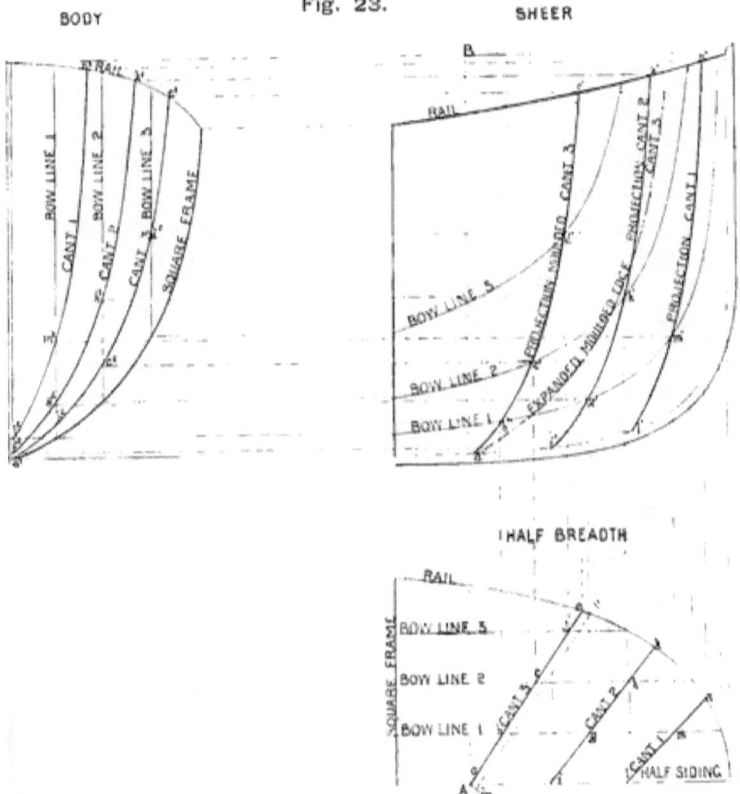

Fig. 23.

keelsons marked upon them. Sometimes a wood mould is made of each with the different markings on and a line for the foot.

Projection of the Cants into the Sheer on the Bow Lines.—Show cants in the Half Breadth and the Body of **Fig. 23** in the manner already described. Square up the points a, b, c, d, and e, until they meet in a^1, b^1, c^1, d^1, and e^1, the corresponding points a^2, b^2,

c^2, d^2, and e^2, produced from the Body, by which No. 3 cant may be drawn in. Repeat the process for cants 1 and 2, and the traces of all the bow lines and cants may be produced.

Expanded Moulded Edge on the Bow Lines.—Produce level lines in the Sheer through points a^1, b^1, c^1, d^1, and e^1, and set off on them from A B the distances A a, A b, A c, A d, and A e on their corresponding level heights. A curve passed through the spots shows the true form of No. 3 cant.

The bevelled edge may be lifted in the same way only along $A^1 p$ from A^1, and set off on the level heights of the intersection of the edge with the bow lines in the Sheer. The manner of terminating at foot and rail has been already indicated.

The Stern Cant Frames.—After the completion of the fairing process of the stern by buttocks and level lines, the cants are arranged around the knuckle line in the Half Breadth, from the transom on one side to the corresponding point on the other, the spacing being the same as the ordinary framing of the square stations. Through these points, on the knuckle, they are drawn as near as practicable square to the rail, knuckle and deck lines, to give the best support to the stern plating. Begin by striking in position the moulded edge of the cants, that is the heel of the bar shown in the Half Breadth of **Fig. 24**, and parallel to it show the bevelled edge. It may be noted that the bosom of the cants at each side of the vessel look towards the centre line to give open bevel.

Projection of the Moulded Edge into the Sheer.—Square up from the Half Breadth the intersection of the moulded edge at a, b, c, and d on to their corresponding lines in the Sheer a^1, b^1, c^1, and d^1. The head of the cant will terminate on the rail at side, d^1, and the foot of the transom will be found by lifting the distance T e in the Half Breadth, and setting it out on the transom in the Body, square to the centre, marked e^1, and then levelling the point over into the Sheer on to the transom station, e^2, as shown by trace lines. A curve is then passed through the points. It will not be a continuous fair curve, for a distinct knuckle is made at b^1. This process is repeated until all the cants are shown in their *true position* in the Sheer.

Projection of the Bevelled Edge into the Sheer.—Square points g, h, k, and l on to their corresponding lines in the Sheer, and transfer T f into the Body, as in the moulded edge, and square over f^1 to f^2, which gives termination of the bevelled edge on the transom.

True Form of the Moulded Edge in the Sheer.—Of course the cants, for the purpose of turning into their proper shape, have to

EXPANSION OF CANTS

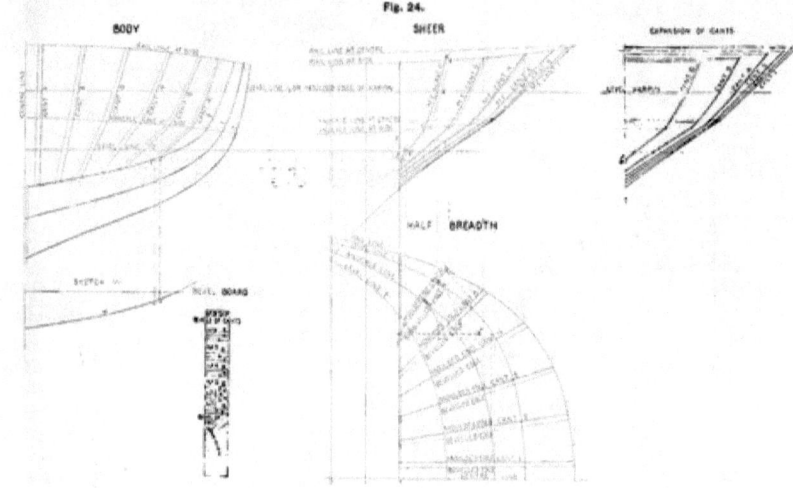

Fig. 24.

be expanded in their *true form*. Place level lines through the projected points b^1 and d^1 on the knuckle and rail. Then lay a batten on the moulded edge of No. 6 cant, and lift the distances a, b, c, and d, from e, and set them out in the Sheer, from the transom, on their corresponding heights or level lines. Through the spots in **Fig. 24** is drawn deep lines, which is the expanded form of the moulded edge of each cant.

True Form of the Bevelled Edge in the Sheer.—Make $e\,a$ square to the moulded edge at the point e, and lift from a on the bevelled edge g, h, k, and l, and set them out from the transom in the Sheer on *their* corresponding heights. The foot is f^2, and the head will be got by drawing a level line through the intersection of the point l, squared on to the rail in the Sheer. Curves are traced through the spots. These lines are shown dotted for distinction.

In this way all the cants are indicated in their true form in the Sheer, or they may be drawn on the loft floor clear of the stern as shown. In many cases these cants are only fitted to the line of the deck, so that may be the sheer height instead of the rail.

Lifting the Bevels.—The amount of bevel is the difference between the expanded moulded and bevelled edges taken square to the moulded edge. Where the bevelled edge lies outside of the moulded edge it is "open," and when inside it is "close." These bevels are lifted at the foot, knuckle, rail, and sometimes intermediate points. They are given to the frame-turner on a piece of pine board the width of the shell flange. The bevels of the feet are also given for cutting the flanges to fit transom floor. The shell flange of No. 6 cant should be cut to a, e, f, seen in the Half Breadth, and the other flange should be cut to b^1, e^2, p.

It may be noted that the bevel at the foot of each cant would be more correctly taken if the distance $e\,a$, the width of the bar, was set out level from e^1 in sketch W, and produced parallel to the centre line to cut T frame; and the point a^1 lifted on to the transom in the expansion in each case, instead of f^2, for the termination of the bevelled edge. This may be understood more clearly by swinging the cant b on the point e, until $e\,d^2$ is square to the transom frame, then a will be on the line e T.

Moulds.—A wood template is made of the expanded moulded edge of each cant, having marked on it the vertical position of the transom, and the position of the rail or deck, and harpin, if any. Where a harpin is fitted a hole should be punched in each cant for attachment. The points where the bevels are taken are also indicated.

be expanded in their *true form*. Place level lines through the projected points b^1 and d^1 on the knuckle and rail. Then lay a batten on the moulded edge of No. 6 cant, and lift the distances a, b, c, and d, from e, and set them out in the Sheer, from the transom, on their corresponding heights or level lines. Through the spots in **Fig. 24** is drawn deep lines, which is the expanded form of the moulded edge of each cant.

True Form of the Bevelled Edge in the Sheer.—Make e a square to the moulded edge at the point e, and lift from a on the bevelled edge g, h, k, and l, and set them out from the transom in the Sheer on *their* corresponding heights. The foot is f^2, and the head will be got by drawing a level line through the intersection of the point l, squared on to the rail in the Sheer. Curves are traced through the spots. These lines are shown dotted for distinction.

In this way all the cants are indicated in their true form in the Sheer, or they may be drawn on the loft floor clear of the stern as shown. In many cases these cants are only fitted to the line of the deck, so that may be the sheer height instead of the rail.

Lifting the Bevels.—The amount of bevel is the difference between the expanded moulded and bevelled edges taken square to the moulded edge. Where the bevelled edge lies outside of the moulded edge it is "open," and when inside it is "close." These bevels are lifted at the foot, knuckle, rail, and sometimes intermediate points. They are given to the frame-turner on a piece of pine board the width of the shell flange. The bevels of the feet are also given for cutting the flanges to fit transom floor. The shell flange of No. 6 cant should be cut to a, e, f, seen in the Half Breadth, and the other flange should be cut to b^1, e^2, p.

It may be noted that the bevel at the foot of each cant would be more correctly taken if the distance e a, the width of the bar, was set out level from e^1 in sketch W, and produced parallel to the centre line to cut T frame; and the point a^1 lifted on to the transom in the expansion in each case, instead of f^2, for the termination of the bevelled edge. This may be understood more clearly by swinging the cant b on the point e, until e d^2 is square to the transom frame, then a will be on the line e T.

Moulds.—A wood template is made of the expanded moulded edge of each cant, having marked on it the vertical position of the transom, and the position of the rail or deck, and harpin, if any. Where a harpin is fitted a hole should be punched in each cant for attachment. The points where the bevels are taken are also indicated.

4

CHAPTER III.

Beam Camber Allowed—Method of Laying it off—To Draw in the Deck at Side Line—Expansion of the Deck Surface—Method of Laying-off Tapered Stringer Plate—Deck Plate Edges—Wide Stringer Plates.

DECKS.

Beam Camber.—The beams of the upper decks of merchant ships are rounded a quarter of an inch for every foot in the length of the beam; that is, a ship 40 feet wide on the midship frame measured level across from the deck at one side to deck at the other side must not have less than 10 inches "camber," or "round up," at the centre line above the level line joining the side line points. This camber is gradually reduced towards the ends, as the beam of the ship decreases. All the beams take the form of the midship camber, so that the deck surface is a fair plane. The other decks when laid follow the same rule. Lower deck beams, when no covering is put upon them, are ordinarily made level. There are several methods adopted for finding the camber curve.

First Method of Laying-off the Camber.—In **Fig. 25**, let A B equal the Half Breadth of the ship at the deck on the midship frame. Set up perpendicular B B^1 equal to the required camber. Join B^1 to A, and erect A A^1 perpendicular to A B^1 at the point A. Produce B^1 to A^1 parallel to B A. Divide B^1 A^1 into a suitable number of equal spaces, say four, A^1 E^1, E^1 F^1, F^1 G^1, G^1 B^1. Divide A B into the same number of equal spaces, A E, E F, F G, and G B. Join E and E^1, F and F^1, G and G^1. Erect A L perpendicular to A B. Divide A L into the same number of equal spaces as A B. Join M, N and O to B^1. Then the cutting points of A^1 B^1 and B B^1, M B^1, and G G^1, N B^1 and F F^1, O B^1 and E E^1, B A and A A^1, give 5 spots for drawing in the curve. Repeat process on the other side of the centre line and you will secure curve for the entire beam.

Second Method.—In **Fig. 26**, let A B equal the length of the beam and C the centre line. Erect C D perpendicular to A B and equal to twice the camber. Join D to A and B. From D on D B set off equal distances, say, one foot apart. Do the same with D A. Number each one, beginning as shown in the sketch, and join 1 to 1, 2 to 2, and so on. It will be seen that these lines form the curve themselves, or the cutting points are a, b, c, d, e, etc.

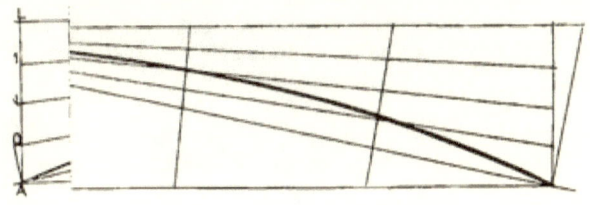

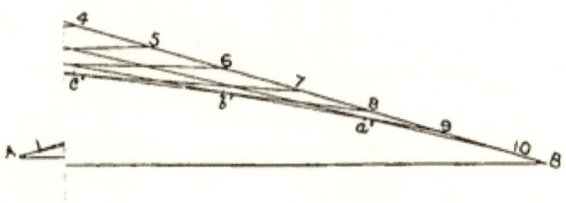

CAMBER FORM.
FIG. 25.

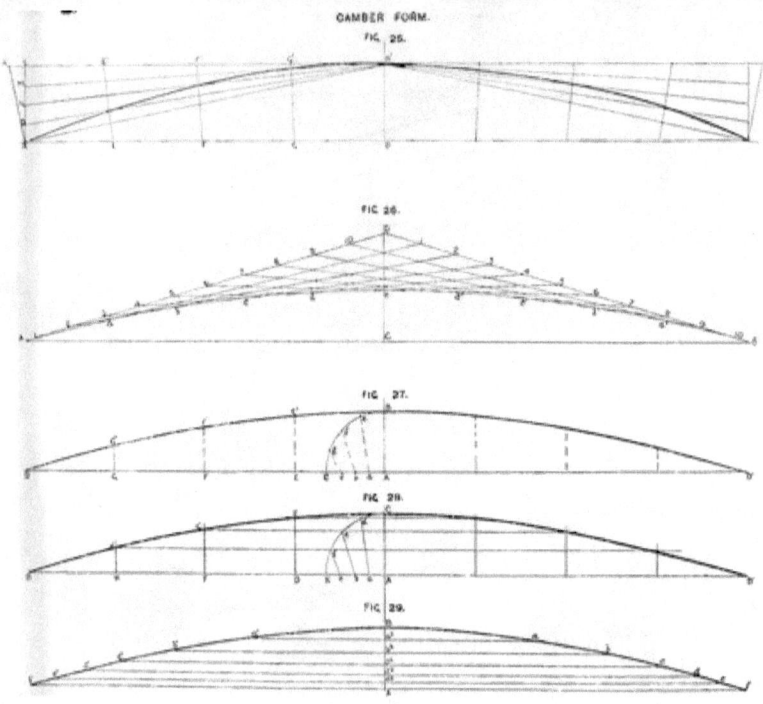

FIG. 26.

FIG. 27.

FIG. 28.

FIG. 29.

Third Method.—In Fig. 27, draw A B perpendicular to A D from A. Describe from A a quadrant of a circle with A C equal to the midship camber. Divide A C into, say, four equal parts by the points a, b, d, and the arc B C into the same number of equal parts by e, f, g, and join a to e, b to f, and d to g. A D equals half beam, divide it into four equal parts in the points E, F, and G. Erect perpendiculars $E E^1 = a e$, $F F^1 = b f$, and $G G^1 = d g$. Draw curve through the points B, E^1, F^1, G^1 and D. Repeat process for the opposite side A D^1, then you have form of full beam.

Fourth Method.—In Fig. 28, let A B = half beam, A C = camber. Describe A C K with radius A C. Divide A K and C K into equal parts and A B into the same number of equal parts as A K, and erect perpendiculars. Produce d to meet E D, e to meet G F, f to meet J H. Pass a curve through the cutting points, repeat for the other side A B^1, and you have an *approximate* method, which is occasionally used.

To Draw in the Deck at Side Line.—Run in the Sheer the deck at the centre, and lift from the Half Breadth the half ordinates of the deck at about every fifth frame. Set these points a, b, c, d, e, and f out square on each side of the centre A B in Fig. 29—the camber curve. Stretch a line across from point to point, a to a^1, b to b^1, and so on, and measure the amount of camber due to the width, B a^2 for section $a a^1$, B b^2 for $b b^1$, etc., and place these distances below the deck at the centre, which are points for the curve of side. Forward the line should terminate on the inside of the stem, aft usually on the centre. At the extreme ends, owing to the rapid reduction of widths, this line will take a sudden rise to the centre line.

Expansion of the Deck Surface.—Girth in the Sheer the deck line at centre for the position of the frame stations, stem and stern points. Lay the batten on a straight line and mark frame stations, etc., from which erect perpendiculars. Set out square to the centre on the camber curve the half beam at each section, and girth the distances of these points around the curve relative to the middle line. Place the girths on their respective perpendiculars in the expansion, and draw curve through the spots. Repeat the curve for the other side, then the enclosed space is a near approximation to the deck area, which may be calculated by Simpson's Rules and the open or uncovered spaces deducted.

To Show a Tapered Side Stringer Plate on the Deck Plan.—In Fig. 30, draw in position the frames 1, 2, 3, 4, and 5, and

Third Method.—In **Fig. 27**, draw A B perpendicular to A D from A. Describe from A a quadrant of a circle with A C equal to the midship camber. Divide A C into, say, four equal parts by the points a, b, d, and the arc B C into the same number of equal parts by e, f, g, and join a to e, b to f, and d to g. A D equals half beam, divide it into four equal parts in the points E, F, and G. Erect perpendiculars $E E^1 = a e$, $F F^1 = b f$, and $G G^1 = d g$. Draw curve through the points B, E^1, F^1, G^1 and D. Repeat process for the opposite side A D^1, then you have form of full beam.

Fourth Method.—In **Fig. 28**, let A B = half beam, A C = camber. Describe A C K with radius A C. Divide A K and C K into equal parts and A B into the same number of equal parts as A K, and erect perpendiculars. Produce d to meet E D, e to meet G F, f to meet J H. Pass a curve through the cutting points, repeat for the other side A B^1, and you have an *approximate* method, which is occasionally used.

To Draw in the Deck at Side Line.—Run in the Sheer the deck at the centre, and lift from the Half Breadth the half ordinates of the deck at about every fifth frame. Set these points a, b, c, d, e, and f out square on each side of the centre A B in **Fig. 29**—the camber curve. Stretch a line across from point to point, a to a^1, b to b^1, and so on, and measure the amount of camber due to the width, B a^2 for section $a a^1$, B b^2 for $b b^1$, etc., and place these distances below the deck at the centre, which are points for the curve of side. Forward the line should terminate on the inside of the stem, aft usually on the centre. At the extreme ends, owing to the rapid reduction of widths, this line will take a sudden rise to the centre line.

Expansion of the Deck Surface.—Girth in the Sheer the deck line at centre for the position of the frame stations, stem and stern points. Lay the batten on a straight line and mark frame stations, etc., from which erect perpendiculars. Set out square to the centre on the camber curve the half beam at each section, and girth the distances of these points around the curve relative to the middle line. Place the girths on their respective perpendiculars in the expansion, and draw curve through the spots. Repeat the curve for the other side, then the enclosed space is a near approximation to the deck area, which may be calculated by Simpson's Rules and the open or uncovered spaces deducted.

To Show a Tapered Side Stringer Plate on the Deck Plan.—In **Fig. 30**, draw in position the frames 1, 2, 3, 4, and 5, and extend them below the centre line. Make $F^1 F^2$ equal to half the width of the stringer plate at the end. Draw $F^1 F$ produced. On No. 1, the half length, set off 1 a the width of the plate at the half length, and on $F^1 F$ the width at the other end, $F^1 b$. Join a to b. Line in centre E F and transfer it on to the plan above, $E^1 F^1$, by making $1^1 E^1$, $2^1 2^2$, $3^1 3^2$, etc., equal to 1 E, 2 e^2, 3 d^2, etc., respectively, and square to the line $E^1 F^1$. Set off on $1^1 a^1$, $2^1 e^1$, $3^1 d^1$, etc., the widths of the plate, 1 a, 2 e, 3 d, etc., and pass curve through the points, which gives the inner edge of the stringer plate. This is the method usually adopted. A more correct one is to girth the line $E^1 F^1$ for the position of the frames and lay them off as shown on the centre line and mark off on the expanded length the width at the end. Join the point with a, shown by a dotted line. Measure the widths at each expanded frame station and mark them off on E^1, 2^2, 3^2, 4^2, and so on. This gives a slightly fuller line. It is, perhaps, in practical work of no great importance.

Iron or Steel Deck Plate Edges.—They are made parallel to the centre line in the expansion plan **Fig. 31**. The side stringer plate inside edge is parallel to the deck at side for the half length amidships, and tapered at the ends in the manner already described. Where no wood deck is laid the strakes are made in and out, but when a wood deck has to be laid (see **Fig. 32**) the strakes are sunken and raised, and the inside edges of the stringer plate may be straight between the butts.

Wide Stringer Plates.—In the case of very wide plates, the inner edge may be arranged parallel to the centre line for half length, and worked in two strakes, shown in **Fig. 33**, then gradually tapered towards the ends. This is the method adopted in war vessels, and there seems no reason why it should not be generally carried out in merchant work.

Fig. 33.

DECK STRINGER PLATE

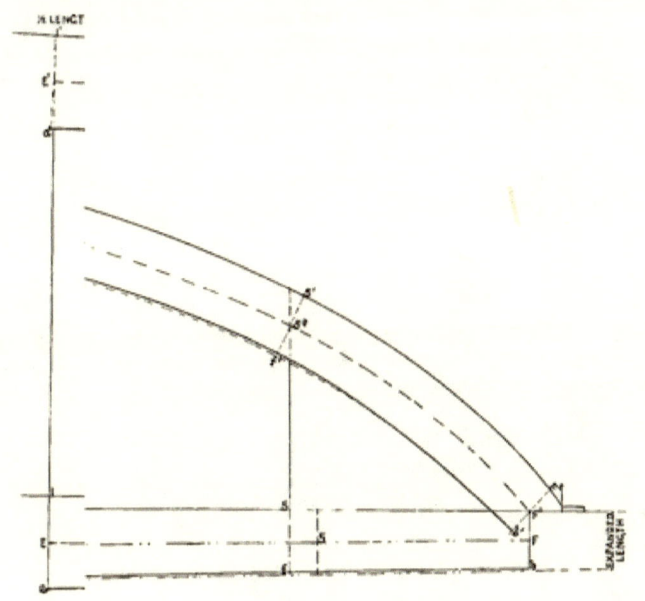

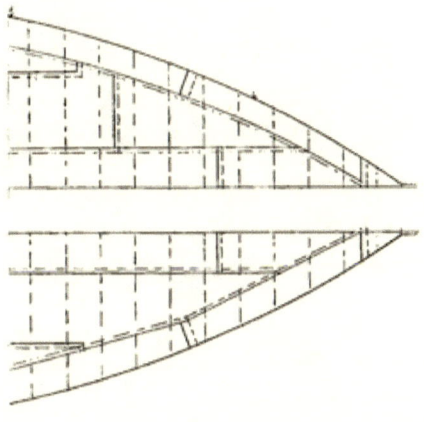

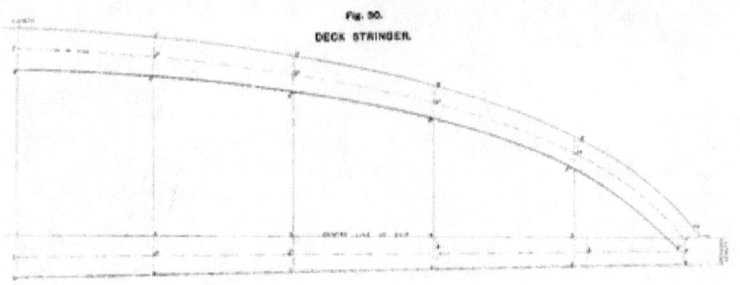

Fig. 30.
DECK STRINGER.

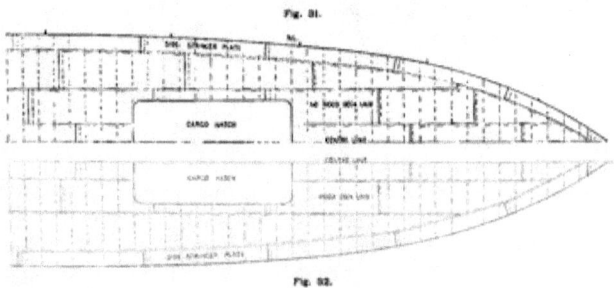

Fig. 31.

Fig. 32.

34.

FLOORS.

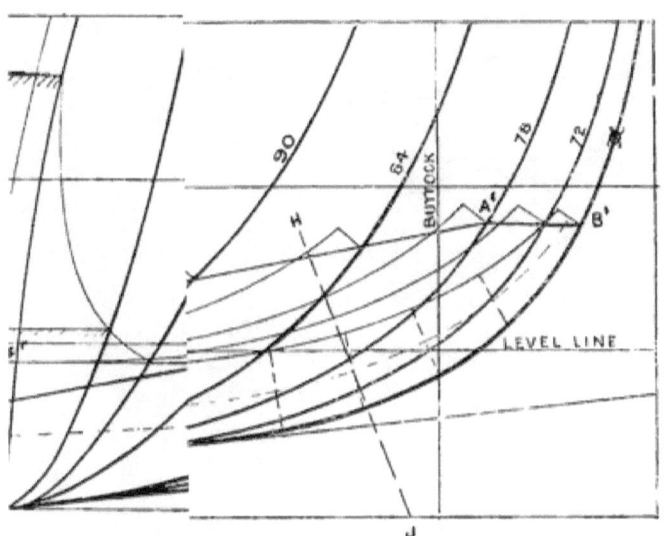

PLAN

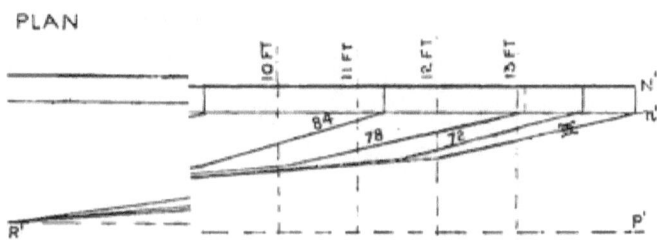

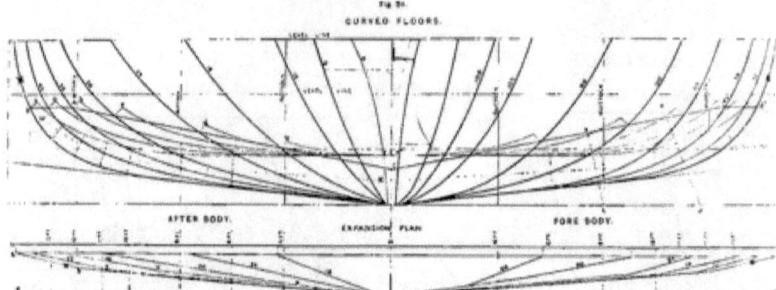

Fig 51.
CURVED FLOORS.

CHAPTER IV.

Turned-up Floors: To obtain the Form—Diminishing Line Fairing-up—Extreme End Floors—Expansion. *Cellular Double Bottoms:* How to Obtain and Fair the Double Bottom—Expansion of the Inner Bottom—Expansion of the Margin Plate—Obtaining Tank Knees—Abaft and Forward of Double Bottom—Expansion of Double Bottom Floors—McIntyre Tank—Swan Conical Tanks.

FLOORS AND DOUBLE BOTTOMS.

Turned-up Floors.—Occasionally ships are built without double bottoms; then the inner edge of the floors are curved up the bilges.

To Obtain the Form of the Curved Floors.—Draw down in the Body, on $\frac{1}{2}$ inch or $\frac{3}{4}$ inch scale, the form of about every sixth frame, extending from the base line to, say, 6 feet up. (See **Fig. 34**.)

The depth of the floor at the centre line is constant, and is settled by the Classification Society's rules. By these rules the midship floor must not be less than half the centre depth, at a distance of three-fourths of the half breadth of the vessel set out from the middle line on the run of the frame, and not less at the extreme ends than the moulding of the frames; and they are required to extend in a fair curve well up the bilges, and in no case to terminate lower at the *outside* of the frame than a perpendicular height of twice the midship depth of the floor above the top of the keel. The figure shows the Body sections in which the height A B is maintained on all floors for one-fourth of the vessel's length amidships, then they are gradually reduced forward and aft until the upper edge of the floors are level with the midships, when they are put in straight. G F is level on all for the centre keelson bars. In ships with exceptionally great rise of bottom the depth at the side should be increased; and those with very flat bottoms, the depth at $\frac{3}{4}$ out should be increased. The reason for this will be evident in laying the curve off.

In the first case, draw in the top edge of the midship floor in both bodies, according to the above rule, and make a skeleton wood mould from the centre outover, marking on it the point E : B E being equal to the moulded frame. Let A B and $A^1 B^1$, on both sides, be the height for one-fourth length amidships. From the point A draw in what is called the "*diminishing line*," cutting the centre line at C,

CHAPTER IV.

Turned-up Floors. To obtain the Form—Diminishing Line—Fairing up—Extreme End Floors—Expansion. *Cellular Double Bottoms*: How to Obtain and Fair the Double Bottom—Expansion of the Inner Bottom—Expansion of the Margin Plate—Obtaining Tank Knees—Abaft and Forward of Double Bottom—Expansion of Double Bottom Floors—McIntyre Tank—Swan Conical Tanks.

FLOORS AND DOUBLE BOTTOMS.

Turned-up Floors.—Occasionally ships are built without double bottoms; then the inner edge of the floors are curved up the bilges.

To Obtain the Form of the Curved Floors.—Draw down in the Body, on $\frac{1}{4}$ inch or $\frac{3}{4}$ inch scale, the form of about every sixth frame, extending from the base line to, say, 6 feet up. (See **Fig. 34.**)

The depth of the floor at the centre line is constant, and is settled by the Classification Society's rules. By these rules the midship floor must not be less than half the centre depth, at a distance of three-fourths of the half breadth of the vessel set out from the middle line on the run of the frame, and not less at the extreme ends than the moulding of the frames; and they are required to extend in a fair curve well up the bilges, and in no case to terminate lower at the *outside* of the frame than a perpendicular height of twice the midship depth of the floor above the top of the keel. The figure shows the Body sections in which the height A B is maintained on all floors for one-fourth of the vessel's length amidships, then they are gradually reduced forward and aft until the upper edge of the floors are level with the midships, when they are put in straight. G F is level on all for the centre keelson bars. In ships with exceptionally great rise of bottom the depth at the side should be increased; and those with very flat bottoms, the depth at $\frac{3}{4}$ out should be increased. The reason for this will be evident in laying the curve off.

In the first case, draw in the top edge of the midship floor in both bodies, according to the above rule, and make a skeleton wood mould from the centre outover, marking on it the point E : B E being equal to the moulded frame. Let A B and $A^1 B^1$, on both sides, be the height for one-fourth length amidships. From the point A draw in what is called the "*diminishing line*," cutting the centre line at C, which is about one-third of the centre depth below the top of the floor. Occasionally this line is curved. The point c depends upon the fineness of the ship's bottom. Set off square at each section, where the diminishing line cuts the frame, the moulded width of the frame B E. Place the point E, on the mould, fair with each of the points E, a, b, c, etc., in succession, and the edge of the mould with F, and chalk in to the mould curve every section floor. Repeat the process in the fore Body: then you have approximately the form of the floors.

Fairing-up the Curved Floors.—The inner edge only requires fairing, so that the surface may be fair for keelsons and ceiling: and this is best done by diagonals like J H and L K, made square to the curves, the distances of the floor edges being lifted from J and L and faired, by the contracted method, in the Sheer plan. The general custom is to use buttocks, lifting the cutting points of the curves on each buttock above the base line and fairing on contracted frame spacing.

The extreme End Floors are level on the top edge and made deeper, like **Figs. 35** and **36**, to form an efficient connection between the sides of the ship, and sufficiently wide between the inside of the frames at the different heights a, b, c, etc., in **Fig. 35**, to fit the keelson bars in. This point should be found for a few floors, and lines $a\,k$ and $a^1\,f^1$ put through the spots and faired-up. Those stepping up the stem and stern post are made deeper to give a good connection. The transom floor must be one and a half times the depth of the midship floor, or the strength made up in some other way. In the case of a screw-steamer, the depth of the forward floors is decided in the same manner as in **Fig. 35**; the after ones must be arranged suitably to stiffen the stern and to clear the stern tube—**Fig. 36** shows the general character of the same with straight tops.

Expansion of the Curved Floors.—Draw down, underneath the Body sections, two parallel lines, N N^1 and P P^1, distance apart the depth of the midship floor at the centre: then girth each floor for its true half length along the middle of its depth, shown by a dotted line $w\,s$ on the midship floor: and set these lengths along M N from M, and drop perpendiculars from the points, like N n, equal to the width of the frame, which is the width of the floors at the ends, and seeing that the ends are all the same width, $n\,n^1$ may be drawn in parallel to M N^1. Set off on $w\,s$ distances about one foot apart, beginning at the centre line: also on M N set the same, and drop perpendiculars from the points. Then lift the width of each floor on

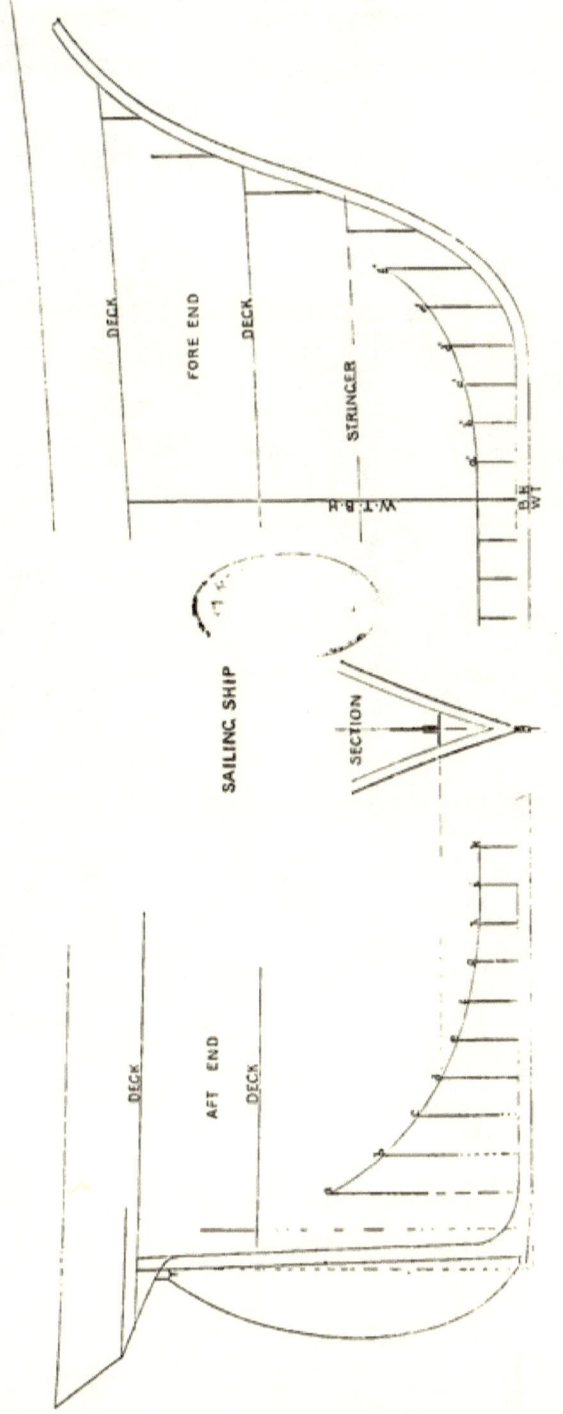

Fig. 36.

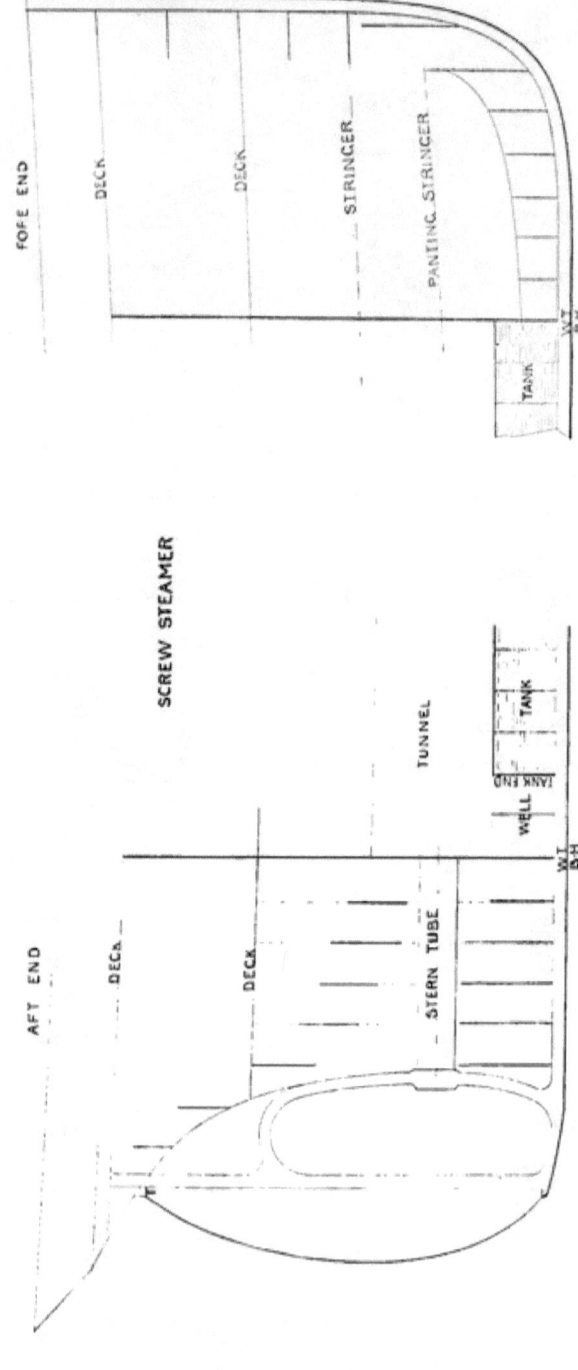

Fig. 36.

SCREW STEAMER

these points square to the dotted line *w x* placed in each section, and lay them off in the expansion plan from the top M N on their corresponding distances from the centre. Make R O the width of the keel bar and parallel to the top. Then a curve passed through each set of spots will give the true form of the plate required. Repeat the process in the fore body. In ordering the plates, the bottom edge is enclosed by straight lines, which will give one or more knuckles on the plate, shown in the after expansion by points S, T, V, X, and R. The intermediate floors are got by dividing the space in between the section points, or by running lines for the top and knuckle edges in the Half Breadth. It is necessary, in ordering from the manufacturer, to give the extreme length on the top edge like M N, also the distances from the centre to the various knuckles with the squared depth thereat. In other words to give sketches of each class of floor with tabulated sizes. Some of the shipbuilders give the actual widths of the floors at about every foot, and allow the mills to cut the lower edge to embrace all these widths : and this seems to be the easiest way for both shipbuilders and millmen, if the widths are taken close and at uniform distances from the centre on every floor, accompanied with one sketch showing the ends, centre, and distances apart of the ordinates. When the floors extend the full width of the ship in one piece, the total lengths must be given with the sketch. If the floors are cut at the centre, two plates will be required for each frame. Sometimes they are butted at different sides on alternate frames about 3 feet away from the centre line, which requires two sketches for each set of floors and dimensions to suit. Allowance to be made for butting if overlapped.

Cellular Double Bottoms.—Most of the screw steamers now built are fitted with double bottoms to carry water ballast, and in some cases large sailing ships are also fitted in this manner. The depth of the double bottom or tank at the centre line should be such that the bottom is easily accessible by man-holes through the floors at all points, and yet allow in way of the man-holes sufficient effective material to sustain probable ordinary strains. Care should be exercised that the man-hole cut out is in the centre of the depth of the floor. If the vessel is built under classification supervision the centre depth ranges from about 32 to 48 inches.

How to Obtain the Lines and Fair the Double Bottom.—When the centre depth A B in **Fig. 37** is settled, a line A C is run parallel to the base line *bb*, showing the inner bottom, which is now made level. Then the width of the flange or margin plate C D is decided upon. This should not be less than 18 inches to get a suitable connection, and need not exceed 34 inches. The width C D is set off square to the midship frame surface at such a point that it touches the inner bottom line C A, and repeated, F G, H J, etc., for about every sixth frame in both bodies as far as the tank extends. The midship width is maintained, where possible, fore and aft; but owing to the ends being much finer it usually has to be gradually reduced from about the three-quarter length. In some cases to get a substantial fastening at fine ends the top is cranked up. After securing these trial lines it is faired by lifting distances of D, F, H, K, etc., square from the centre, and placing them in the Half Breadth on their respective frames; any correction for fairness made in the Body. Then the points C, G, J, L, etc., are lifted from A and also placed in the Half Breadth and faired. The intermediate spots are lifted from these lines for the scrieve board. The fore body sections are treated in the same manner for the extent of the inner bottom. Of course the after and forward body lines should be continuous curves. Some shipbuilders allow A C to drop slightly towards the point C, if the taper on the tank floors can be brought within 9 inches.

Having now obtained the lines and faired them, you proceed to make expansions of the various parts for ordering the plating.

Expansion of the Inner Bottom Plating.—A base line is drawn representing the length of the top on a $\frac{1}{4}$ inch scale. Perpendiculars are erected at the frame stations. The widths on the top from the centre A to the knuckle line C, G, etc., at each frame are lifted and set off in this expansion on their corresponding stations, and a curve passed through the spots. Then the breadths of the margin plate from the knuckle to the frame, C D, G F, etc., are lifted and laid out from the knuckle line and the curve drawn. A line not less than $1\frac{1}{2}$ inches parallel is made, inside of this knuckle, representing the inner edge of the margin plate, a in **Fig. 38**, when flanged to take the double bottom plating, which is attached to it with a single riveted lap. Next set off parallel to the centre line the position of the intercostal girders between the floors, and arrange top plating edges parallel to the centre clear of these. All the frames should now be drawn across the plan. Show the engine seating plating, position of the bunkers, boiler bearers, engine girders, man-holes, bulkheads, tunnel, shell plating butts, and whatever has to come in contact with the inner bottom plating; after which the butts may be arranged and the plates

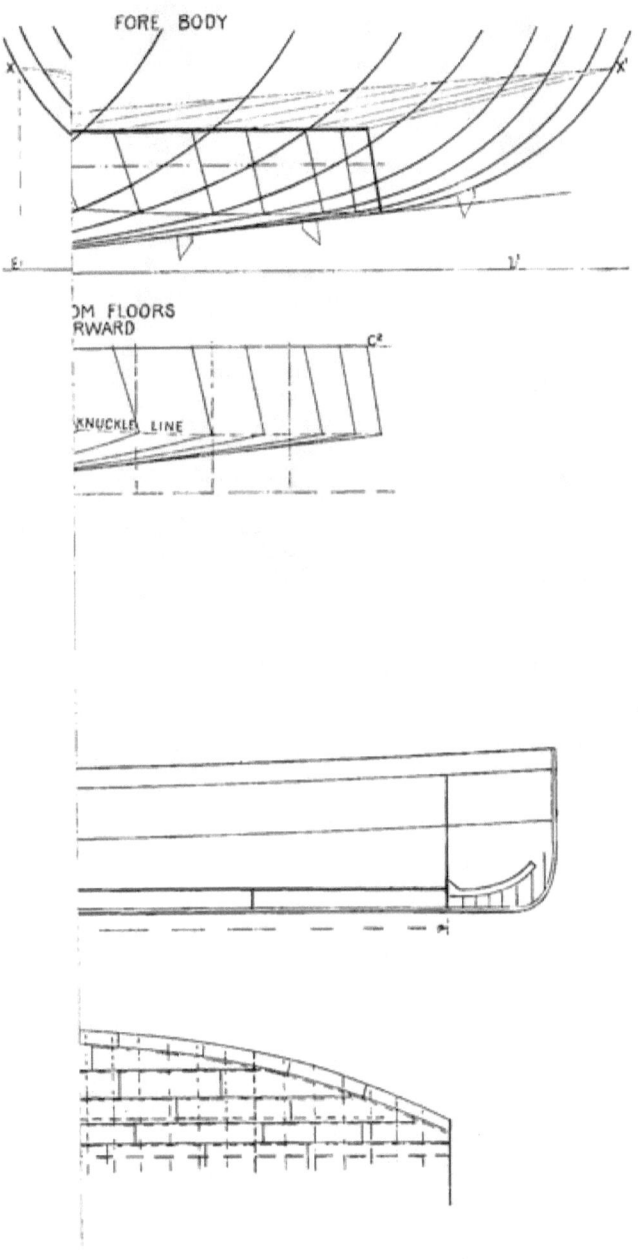

Fig. 37.
CELLULAR DOUBLE BOTTOM.

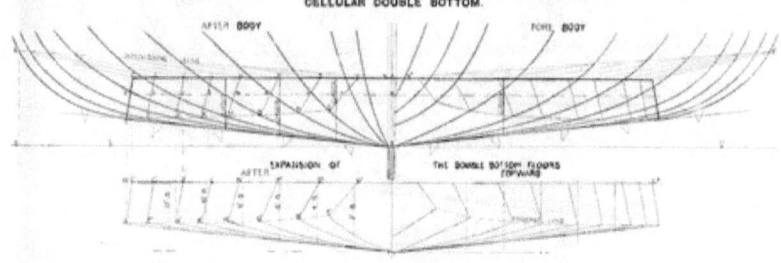

Fig. 39.
SHEER.

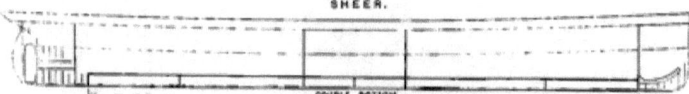

EXPANSION OF INNER BOTTOM

ordered. It is not a wise plan to measure the lengths from the drawing with a scale. A much safer way is to calculate the length of tank top, and add the amount required for butts, and then work in plates as near as possible of the same length—an allowance of $\frac{1}{4}$ to $\frac{1}{2}$ an inch on each plate is sufficient. The margin plate is made straight on its inner edge from butt to butt. It will be evident that where the margin plate is flanged at the knuckle, a little less in the width of plate would do, about $\frac{1}{2}$ an inch, owing to the knuckle being part of a circle, this deduction can be easily made when ordering. It will also be readily seen that this method of expanding the margin plate, although generally used, is on the big side, which has led some shipbuilders in these days of economy to get it more accurately in the following manner:—

Expansion of the Margin Plate.—Place the level line $g\ o$ in the Body, **Fig. 37**. Lift points g, h, c, l, m, etc., from the centre line, and place them in the Half Breadth on their respective frames. Then girth this line for the position of the frames and place these, expanded, on a straight line about the midship frame, from which points square lines indefinitely above and below. Lift the widths of the margin plate below the level line on the Body sections g D, h F, etc., and transfer them to their respective expanded frame stations below the level line. Pass a curve through the spots which will give the lower edge of the margin plate; allowance may be made for the point b, in **Fig. 38**, being about $\frac{1}{2}$ inch off the shell. The top edge may be found on a large section, or on the loft floor Body, by drawing the curve of the plate and its position on the inner edge (see **Fig. 38**). Then girth from g, h, i, etc., to each inner point a and set these girths off above the expansion line $g\ o$ on their respective stations, and draw a curve through the spots. This is the expanded top edge of the margin plate. The process is repeated for the fore body. The form of this margin plate may also be found in the same way as the longitudinal explained on page 132.

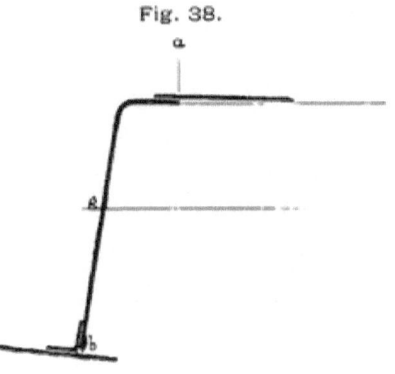

Fig. 38.

ordered. It is not a wise plan to measure the lengths from the drawing with a scale. A much safer way is to calculate the length of tank top, and add the amount required for butts, and then work in plates as near as possible of the same length—an allowance of ¼ to ½ an inch on each plate is sufficient. The margin plate is made straight on its inner edge from butt to butt. It will be evident that where the margin plate is flanged at the knuckle, a little less in the width of plate would do, about ½ an inch, owing to the knuckle being part of a circle, this deduction can be easily made when ordering. It will also be readily seen that this method of expanding the margin plate, although generally used, is on the big side, which has led some shipbuilders in these days of economy to get it more accurately in the following manner:—

Expansion of the Margin Plate.—Place the level line $g\ o$ in the Body, **Fig. 37**. Lift points g, h, e, l, m, etc., from the centre line, and place them in the Half Breadth on their respective frames. Then girth this line for the position of the frames and place these, expanded, on a straight line about the midship frame, from which points square lines indefinitely above and below. Lift the widths of the margin plate below the level line on the Body sections g D, h F, etc., and transfer them to their respec-

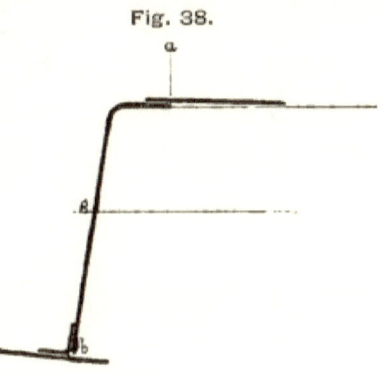

Fig. 38.

tive expanded frame stations below the level line. Pass a curve through the spots which will give the lower edge of the margin plate; allowance may be made for the point h, in **Fig. 38**, being about ½ inch off the shell. The top edge may be found on a large section, or on the loft floor Body, by drawing the curve of the plate and its position on the inner edge (see **Fig. 38**). Then girth from g, h, i, etc., to each inner point o and set these girths off above the expansion line $g\ o$ on their respective stations, and draw a curve through the spots. This is the expanded top edge of the margin plate. The process is repeated for the fore body. The form of this margin plate may also be found in the same way as the longitudinal explained on page 132.

Obtaining Tank Knees.—The tank knees are next drawn in. The point E X on the midship frame is fixed as per the Classification Society's rules, and a diminishing line X Y is drawn in each Body for the height of all knees on the frame edge. This line should be at the after and forward extremities sufficiently high to get a good connection for the knee. Then the points C, G, J, etc., are joined to the "diminishing line" at their respective frames. Occasionally the knees are hollow between these extreme points, and sometimes the knee is kept about $\frac{3}{4}$ inch below the points C, G, etc.

Abaft and Forward of the Double Bottom.—In **Figs. 39** and **36** the extent of the inner bottom is shown, and the height of the floors indicated before and abaft of it. These are ordered in the usual way, being all straight on the top edge.

Expansion of the Double Bottom Floors.—Draw down under **Fig. 37**, level line $C^1 C^2$ for the top of the floors, on to which square down points C, G, J, etc., and trace form of each section in as shown, which may be easily done on close buttock lines shown dotted in the expansion. In ordering enclose the curved bottom edge by straight lines, giving a series of knuckles. Some give the depth from the top edge to the curve on close spaced buttocks, and allow the mills to sheer the plate edge, between the buttocks, which is very small. The depth at and outstretch from the centre must be given at D W. The intermediate floors are found by "running" the top and knuckle points in the Half Breadth. Those on the buttocks may be found by "inspection."

NOTE.—In a few cases the old McIntyre tank is still fitted. The tank side and top are the same as the cellular bottom, only the floors are shallow and straight. Continuous fore and aft girders being fitted between the floors and the inner bottom. The only difference, as far as laying off is concerned, is that the floors are straight and level on the top edge. The Swan conical tank has been fitted in some cases; but, seeing the purpose for which it was invented has passed out of use, it is not described here.

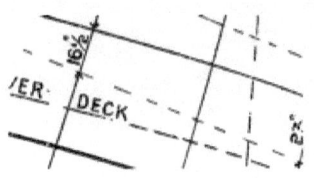

CHAPTER V.

Shell Plating; Obtaining the Sight Edges – Lining Off the Model – Fairing the Sight Edges on the Model, on the ¾ lines, on the Loft Floor – Transferring the Sight Edges to the Scrieve Board – Ordinary Shell Expansion – Area of the Outer Bottom – Ordering the Shell Plating – Stern Expansion, obtaining True Form of Plating, Check on the Expansion – Another Method of Stern Expansion – Tumble-home Stern Expansion.

SHELL PLATING.

Obtaining the Sight Edges.—Those are called sight edges which are visible on the outside, and they are first obtained on the office Sheer Draught in the following manner :—Mark off on the Body plan midship section in **Fig. 40**, on each side, the sight edges of the sheer and garboard strakes. It is usual to place the top edge of the sheer strakes from 10 to 12 inches above the upper deck moulded line. The garboards are flanged down unto the bar keel to within half an inch of the bottom of the keel. These two points fixed, divide the sight edges of the intermediate strakes in equal or suitable widths, or lift the widths from the ½ inch "Scantling Section," on which they are commonly arranged to suit the Classification Society's rules. Amidships, for about the half length, above the light draft line, the edges are sheered to the upper deck, W Y being midship width, and from this point in the after body they are run square to the frame curve to reduce the amount of curvature or "sny" on the plates. In the fore body from about the half length to the stem they may be reduced in width from 1 to 3 inches, and drawn in a fair curve. Those below the light line, in both bodies, should be sketched in about parallel girths for the half length, and then run in straight to the stem and stern posts about square to the frame curve. Many shipbuilders make all the lower edges straight in the Body. It is best, if you can, to arrange the strake taking the screw boss in one plate and an outside strake. This cannot always be done, because of the plate being too large for the furnace, and other points of consideration of more importance to the shipbuilder. Owing to the girth of the frames decreasing considerably towards the ends, it is requisite to stop one or more of the edges some distance from the stem and stern. A strake of this description is called a "lost strake" or stealer. **Fig. 40** shows the plate edges of a screw steamer somewhat disposed on this plan.

CHAPTER V.

Shell Plating: Obtaining the Sight Edges—Lining Off the Model—Fairing the Sight Edges on the Model, on the ⅜ lines, on the Loft Floor—Transferring the Sight Edges to the Scrieve Board—Ordinary Shell Expansion—Area of the Outer Bottom—Ordering the Shell Plating—Stern Expansion, obtaining True Form of Plating, Check on the Expansion—Another Method of Stern Expansion—Tumble-home Stern Expansion.

SHELL PLATING.

Obtaining the Sight Edges.—Those are called sight edges which are visible on the outside, and they are first obtained on the office Sheer Draught in the following manner:—Mark off on the Body plan midship section in **Fig. 40**, on each side, the sight edges of the sheer and garboard strakes. It is usual to place the top edge of the sheer strakes from 10 to 12 inches above the upper deck moulded line. The garboards are flanged down unto the bar keel to within half an inch of the bottom of the keel. These two points fixed, divide the sight edges of the intermediate strakes in equal or suitable widths, or lift the widths from the ½ inch "Scantling Section," on which they are commonly arranged to suit the Classification Society's rules. Amidships, for about the half length, above the light draft line, the edges are sheered to the upper deck, W Y being midship width, and from this point in the after body they are run square to the frame curve to reduce the amount of curvature or "sny" on the plates. In the fore body from about the half length to the stem they may be reduced in width from 1 to 3 inches, and drawn in a fair curve. Those below the light line, in both bodies, should be sketched in about parallel girths for the half length, and then run in straight to the stem and stern posts about square to the frame curve. Many shipbuilders make all the lower edges straight in the Body. It is best, if you can, to arrange the strake taking the screw boss in one plate and an outside strake. This cannot always be done, because of the plate being too large for the furnace, and other points of consideration of more importance to the shipbuilder. Owing to the girth of the frames decreasing considerably towards the ends, it is requisite to stop one or more of the edges some distance from the stem and stern. A strake of this description is called a "lost strake" or stealer. **Fig. 40** shows the plate edges of a screw steamer somewhat disposed on this plan.

Lining Off the Model.—A half block model of the ship is made on a scale of a quarter of an inch equal to 1 foot. The back is slightly beyond the centre line as shown in **Fig. 41**, for marking off on the keel the frame stations. The top is flush to the rail line or poop and forecastle decks, as the case may be. This model is painted white and smoothed down, or varnished. The top and back are not painted.

Fig. 41.

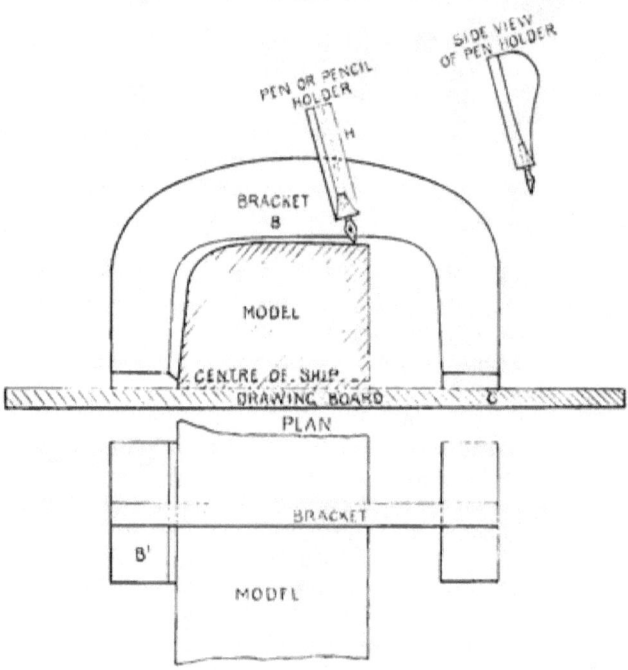

The upper deck line should be set off on the midship frame, and then run in fore and aft with a gauge parallel to the top line. If not parallel it is lined in with a batten attached with pins or eyeless needles. The model is laid upon a thick drawing board C, and the position of the frames marked on the bottom centre by producing the model maker's section lines from the back, and setting off the intermediates. This producing is done with a joiner's set square. The stations are also squared across the top from the back to act as a guide in lining in the frames; which is done with a drawing-pen and ink, or with a pencil, by placing brackets B of various form across the model to guide the pen or pencil holder H, made with a flat side to slide easily and steadily against the face of the bracket. The palm edge B^1 is perfectly square to the face B, and is worked close up against the keel, and so fixed to the frame station that the pen will trace through the spot. When all the frames are drawn across the model, the position of the shell plate sight edges are lifted, relative to the upper deck and centre line, on narrow strips of drawing paper. This is done by girthing the body sections (on the same scale as the model), and marking centre, plate edges, and decks, and transferring the girthed spots to the model on corresponding frames. Thin narrow yellow pine battens of parallel width are pinned to the spots, and the lines faired and inked or pencilled in, and corrections, if any, made in the Body. The battens may be kept in position by strong elastic bands passed round the model. The stringers, keelsons, and inner bottom side are lined off in the same way. Then the butts of the shell plating, stringers, keelsons, etc., are transferred from the "shell expansion," or arranged by the draftsman on the model. It is usual to line-off all wash ports, scuppers, sidelights, openings in the shell, erection fronts, bulkheads, liners, doublings, alphabetical letter of strake, number of plate, thickness of plate; and to indicate, by various colours, treble, double, and single riveted butts. Many shipbuilders show the character of butts by cross lines, one line standing for single, two for double, and three for treble. The shell plating is usually ordered from this model, so that it is necessary that the marking off should be done with care and completeness.

Fairing Plate Edges on ¾ inch Lines.—In this case the plate edges are obtained on the ¾ inch scale Body sections in the manner already described. Those above the light draught line are faired-up in the Sheer by lifting at each sectional frame the perpendicular distance like T S from the base to the intersection of the plate edge and frames, and setting the heights above the base in the Sheer on their corresponding frames. Those below the light line are lifted square out, like O P, from the centre line to the frame and plate edge intersection and laid-off in the Half Breadth. Of course the correct terminations must be found by projecting the cutting points, as in n, on the half siding into the Sheer on the inside of the posts, and from there into the Half Breadth. Any modification necessary for fairness is made in the Body.

Fairing Plate Edges on the Loft Floor.—Carry out the same method as explained for the ¾ inch lines.

Transferring Shell Sight Edges to the Scrieve Board.—The position of the sight edges on a few sections may be girthed and given to the loftsman, but what appears a better plan is to draw straight lines from the post to the midship point, like $a b$, $c d$, $e f$, and $g h$, in Fig. 40, and girth the distance on the frames from these lines to the sight edge. If the loftsman is supplied with a copy of these sections with the lines and distances on, and the vertical height on the midship section frame and centre line, he ought to have little or no difficulty in transferring. Of course, consideration should be given to any alteration found necessary in the scrieve midship widths.

The Ordinary Method of making a Shell Expansion.—In Fig. 42, set off to a quarter inch scale the length of the ship on a base line T F, and lift the form of the after end, A B C, and the fore end, D E F, from the Sheer Draught; or lay the model on the base and pencil round the ends. Mark off on the base the position of the Body sections, and erect from these points perpendiculars of indefinite length. Then girth each Body section, drawn to the same scale, from the centre line to the rail, for the position of the plate edges, keelsons, decks, etc., and lay these distances out on their respective stations from the base T F. The terminations of these lines on the inside of the stem and stern posts are lifted perpendicular from the base of the Body on the half siding, and transferred on to the expansion. The shape of the part abaft of the transom is taken from the stern expansion. Lines may now be run through the spots, as shown in Fig. 42, and all the frames drawn in red ink to their correct heights. The transverse bulkheads, decks, keelsons, tank side, and girders are indicated by blue lines, and the shell plating edges, butts, openings, in black. Doubling plates should also be shown in some distinguishing colour. This plan shows all openings in the bottom, and whatever comes in contact with it, besides the alphabetical mark of each strake, and the number of each plate, the size and spacing of the riveting, the width of the butt straps or laps, and the seam edge laps, and at least the half and three-quarter's length amidships, and the change of the frames or deadflat. It will be evident that this is not a correct expansion of the shell plating; and is, therefore, of no use for measuring off the size of the plates. Its purpose is solely for the *guidance* of the shell men in plating the bottom of the ship, which decides the information to be placed upon it.

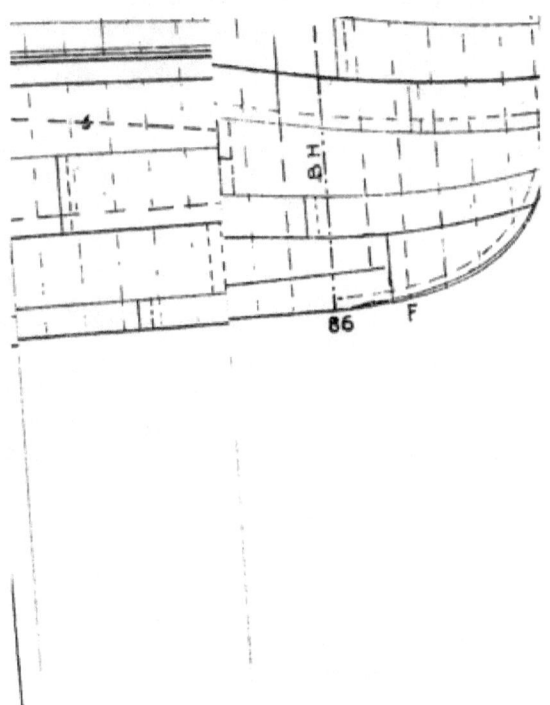

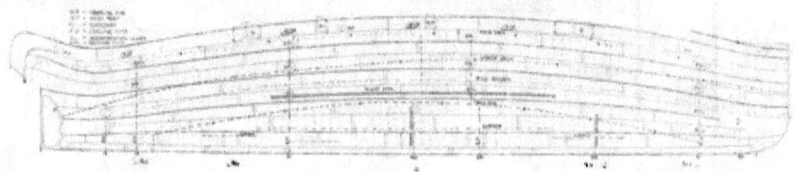

FIG. 42
SHELL EXPANSION PLAN

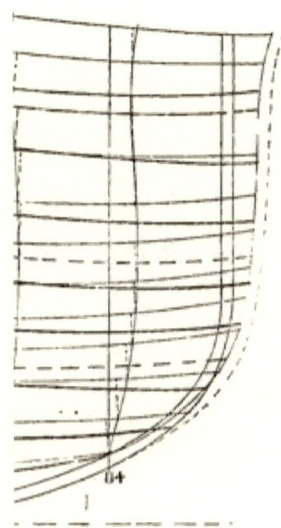

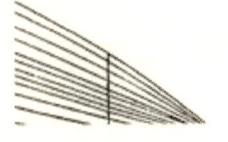

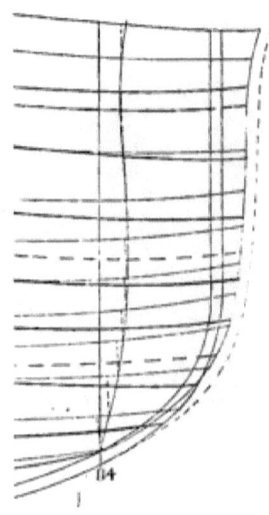

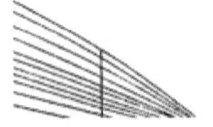

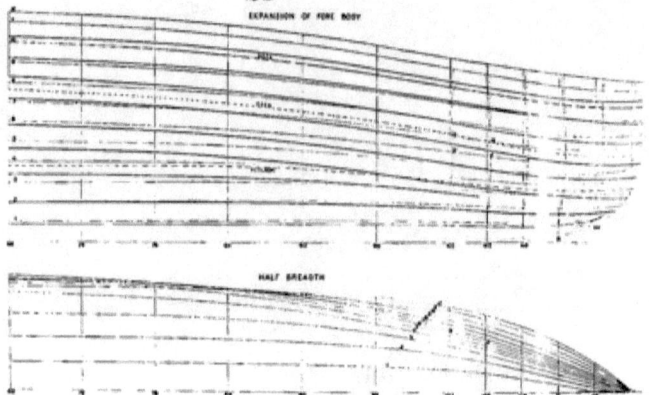

Fig. 49.
EXPANSION OF FORE BODY

HALF BREADTH

Area of the Outer Bottom or Shell Plating.—An approximate expansion of the shell plating may be found by taking the lengths of a series of level lines, and of a series of transverse sections, and manipulating their intersections by two sets of paper strips. Pinning the strips to the centre line position, with freedom to move round that point, and connecting the cutting points of the transverse strips with the longitudinal strips, also free to move, by this means the surface is laid out approximately in a level plane. But this is not correct in principle, owing to no account being taken of the inclination of the level lines to the vertical sections. A more correct method is by a writer in one of the Annuals of the Royal School of Naval Architecture, signing himself J. C. For instance, in **Fig. 43** let $a\ b$ and $d\ c$ be two sections. Draw $b\ c$ and $a\ d$ as near as possible square to both sections 105 and 102. Then these lines, when placed on the model, will be almost perpendicular to each other. In order to find the area of the space enclosed, it will only be necessary to obtain the lengths of $b\ c$ and $a\ d$, and then construct the figures thus. In the expansion make A D = the true length of $a\ d = A^1\ D^1$. Erect perpendicular D C and A B to A D line, and draw A B = $a\ b$ and D C = $d\ c$, and join B to C. This figure A, B, C, D, represents closely the area of the surface on the model. Apply this principle to the entire surface. Draw in the Body curved diagonal lines, 1, 2, 3, etc.—shown faint lines—as near as possible square to the frames at their points of intersection. Then girth from the centre the position of the frames on these lines, and set them off in the Half Breadth as shown. Secure the end terminations by squaring over into the Sheer the heights on the inside of the stem, and projecting down into the Half Breadth on the half siding. By this means the true lengths on the diagonals are got. Then apply these, beginning with the bottom, and set off on the stations in the expansion the square breadths or girths of the lowest. Draw lines through the spots thus obtained, and apply the girthed positions of the frames taken from the Half Breadth on this first diagonal, commencing from number 66 and working therefrom. Erect perpendiculars, in the expansion to the curve of first diagonal from the points, and apply the widths girthed from the Body between one and two diagonals, and run curve through the spots. The position of the frames for number two diagonal, taken from the Half Breadth, must be set on the line just drawn in, and the girthed widths from the Body set on the perpendiculars as before, which will give spots for another expanded diagonal line. The process is

Area of the Outer Bottom or Shell Plating.—An approximate expansion of the shell plating may be found by taking the lengths of a series of level lines, and of a series of transverse sections, and manipulating their intersections by two sets of paper strips. Pinning the strips to the centre line position, with freedom to move round that point, and connecting the cutting points of the transverse strips with the longitudinal strips, also free to move, by this means the surface is laid out approximately in a level plane. But this is not correct in principle, owing to no account being taken of the inclination of the level lines to the vertical sections. A more correct method is by a writer in one of the Annuals of the Royal School of Naval Architecture, signing himself J. C. For instance, in **Fig. 43** let $a b$ and $d c$ be two sections. Draw $b c$ and $a d$ as near as possible square to both sections 105 and 102. Then these lines, when placed on the model, will be almost perpendicular to each other. In order to find the area of the space enclosed, it will only be necessary to obtain the lengths of $b c$ and $a d$, and then construct the figures thus. In the expansion make A D = the true length of $a d$ = $A^1 D^1$. Erect perpendicular D C and A B to A D line, and draw A B = $a b$ and D C = $d c$, and join B to C. This figure A, B, C, D, represents closely the area of the surface on the model. Apply this principle to the entire surface. Draw in the Body curved diagonal lines, 1, 2, 3, etc.—shown faint lines—as near as possible square to the frames at their points of intersection. Then girth from the centre the position of the frames on these lines, and set them off in the Half Breadth as shown. Secure the end terminations by squaring over into the Sheer the heights on the inside of the stem, and projecting down into the Half Breadth on the half siding. By this means the true lengths on the diagonals are got. Then apply these, beginning with the bottom, and set off on the stations in the expansion the square breadths or girths of the lowest. Draw lines through the spots thus obtained, and apply the girthed positions of the frames taken from the Half Breadth on this first diagonal, commencing from number 66 and working therefrom. Erect perpendiculars, in the expansion to the curve of first diagonal from the points, and apply the widths girthed from the Body between one and two diagonals, and run curve through the spots. The position of the frames for number two diagonal, taken from the Half Breadth, must be set on the line just drawn in, and the girthed widths from the Body set on the perpendiculars as before, which will give spots for another expanded diagonal line. The process is repeated until the surface is completed; when it may be calculated by Simpson's rules, the butts and seams being got by percentage. The frames, dotted, appear curved lines. The position of the decks, keelsons, plate edges, may be transferred with reference to the diagonals. In **Fig. 43**, the decks and keelsons are dotted lines and the plate edges thick lines.

Ordering of the Shell Plating.— The nett widths at the butts of each strake are lifted from the scrieve board. This is done by marking on the "boards," in chalk, the position of the butts between the frames, and girthing the line in each case for the *full* width of the strake, *i.e.*, including the laps. The draftsman makes an allowance in ordering for the "say," or round on the edge, which is practically a matter of judgment. A rough idea can be got from the model by holding a scale along the plate edge. One of the best methods is to place a piece of thin tracing paper on the model, covering the plate lines from butt to butt, and tracing carefully with a pencil the top and bottom edges A B C and F E D, and the butts A F and C D, see **Fig. 44**; then straight lines are drawn enclosing this trace. Draw A C and d e to enclose plate, and bisect the space by a b, make d a f and C b e square to a b; then d e C f is the nett size of the plate required of double taper, to which must be added from half an inch to an inch in the length for planing. The allowances on the breadths should be very little at midships in the case of the inside strakes, and from three-quarters to one inch on the outside strakes. It is usual to allow a few inches in the length of the endmost plates, and also a greater allowance on the widths of the plates under the counter, to meet any alteration made to the plate edges on the ship. A necessary precaution, when the ship is scrieved down, is to girth a few sections and test the correctness of the model.

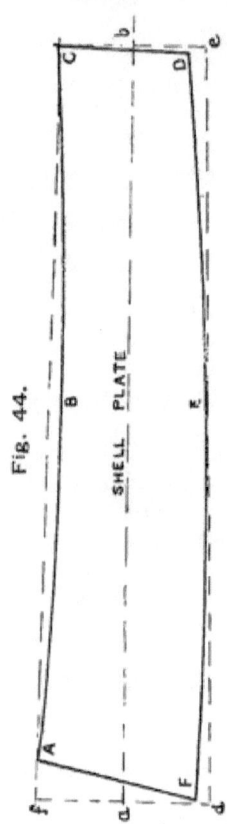

Fig. 44.

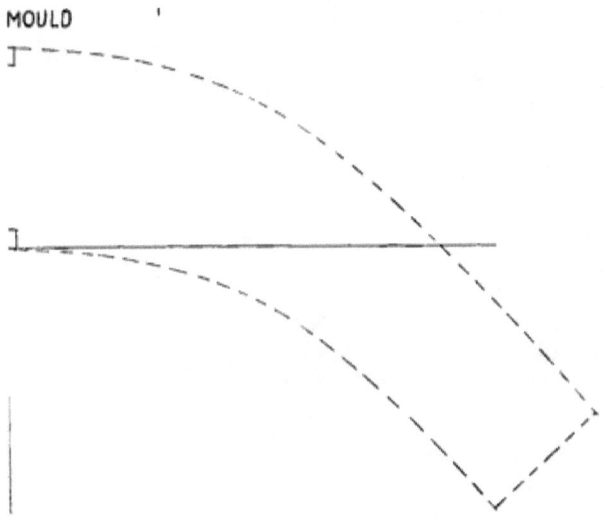
MOULD

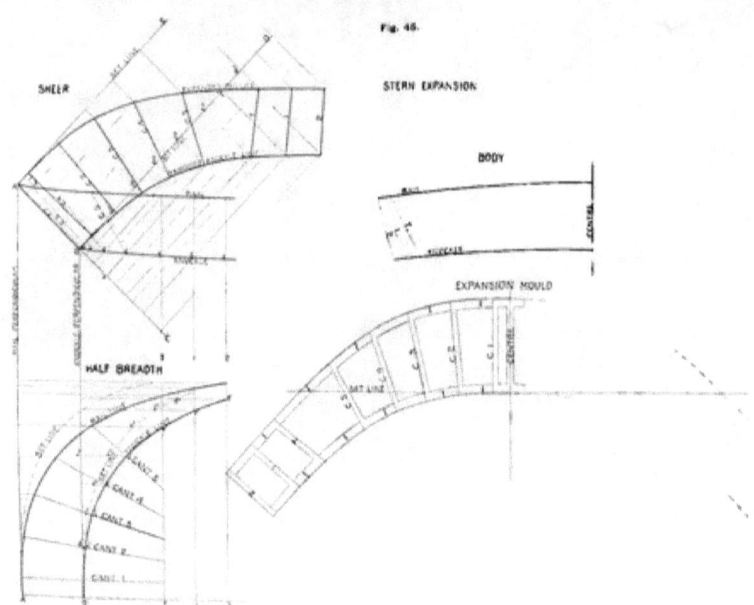

Fig. 46.

STERN EXPANSION.

How to Obtain the True Form of the Stern Plating.—In Fig. 45. produce A B indefinitely to C, and erect square set lines B D from B, and A E from A. Join A to A¹ and B to B¹. Through the points a^1, b^1, c^1, d^1, e^1, f^1, and g^1 in the Half Breadth, draw level lines to the knuckle perpendicular, and drop perpendiculars from a, b, c, d, e, f, and g in the Sheer on to A C. Lift from A C the length of these lines from the knuckle, and set them forward of the knuckle perpendicular B B¹, on their corresponding level lines in the Half Breadth: for instance, J d on J¹ d^1. Pass a curve through these spots, and you have what is called the corrected knuckle or "set line:" girthing it from the centre for the points of the cants and frames, e^2 will be the mark for the transom, and so on. Lay them out in the Sheer from B on B D, and erect perpendiculars of an indefinite length at the points a^2, b^2, etc., and then produce lines parallel to B D through a, b, c, etc., to meet corresponding perpendiculars: a should cut a^2, and so on. A curve passed through the intersections is the expanded knuckle line. The same process is repeated for the rail, working from A A¹ and A E: by which the true moulded form of the plating is got. Sometimes an additional set line is put in between the rail and knuckle. Where there is much round on the transom, it is a wise plan to girth the distance from the knuckle to the rail in the Body, and see that the expansion is sufficiently wide.

Check on the Stern Expansion.—Bend a lath round the original rail and knuckle lines in the Half Breadth, and lift the position of the cants and ordinary frames relative to the centre line, which try on the respective expansion lines of the rail and knuckle: if correct, the position of the cants, etc., should agree.

The butts of the plating are now arranged clear of the cants, etc., and the plates ordered, making suitable allowance at the top and bottom for connections. A wood template is made of the expanded edges, from the centre line to the butt just forward of the transom. Cross-pieces are nailed on it, for the moulded edge of the cants and frames, over which is marked the position of the set line B D. The other side of the ship is simply duplicated. The plater makes the allowances for connection at top and bottom over and above the template. A mould is also given to the yard showing the shape of the set line.

Another Method of Stern Expansion.—In Fig. 46, erect set line A B square to A C from the point A, and square down on to the cants in the Half Breadth the intersections of this set line with the rail

STERN EXPANSION.

How to Obtain the True Form of the Stern Plating.—In Fig. 45, produce A B indefinitely to C, and erect square set lines B D from B, and A E from A. Join A to A¹ and B to B¹. Through the points $a^1, b^1, c^1, d^1, e^1, f^1$, and g^1 in the Half Breadth, draw level lines to the knuckle perpendicular, and drop perpendiculars from a, b, c, d, e, f, and g in the Sheer on to A C. Lift from A C the length of these lines from the knuckle, and set them forward of the knuckle perpendicular B B¹, on their corresponding level lines in the Half Breadth: for instance, J d on J¹ d^1. Pass a curve through these spots, and you have what is called the corrected knuckle or "set line:" girthing it from the centre for the points of the cants and frames, c^2 will be the mark for the transom, and so on. Lay them out in the Sheer from B on B D, and erect perpendiculars of an indefinite length at the points a^3, b^3, etc., and then produce lines parallel to B D through a, b, c, etc., to meet corresponding perpendiculars: a should cut a^3, and so on. A curve passed through the intersections is the expanded knuckle line. The same process is repeated for the rail, working from A A¹ and A E: by which the true moulded form of the plating is got. Sometimes an additional set line is put in between the rail and knuckle. Where there is much round on the transom, it is a wise plan to girth the distance from the knuckle to the rail in the Body, and see that the expansion is sufficiently wide.

Check on the Stern Expansion.—Bend a lath round the original rail and knuckle lines in the Half Breadth, and lift the position of the cants and ordinary frames relative to the centre line, which try on the respective expansion lines of the rail and knuckle: if correct, the position of the cants, etc., should agree.

The butts of the plating are now arranged clear of the cants, etc., and the plates ordered, making suitable allowance at the top and bottom for connections. A wood template is made of the expanded edges, from the centre line to the butt just forward of the transom. Cross-pieces are nailed on it, for the moulded edge of the cants and frames, over which is marked the position of the set line B D. The other side of the ship is simply duplicated. The plater makes the allowances for connection at top and bottom over and above the template. A mould is also given to the yard showing the shape of the set line.

Another Method of Stern Expansion.—In Fig. 46, erect set line A B square to A C from the point A, and square down on to the cants in the Half Breadth the intersections of this set line with the rail and cants, a, b, c, and d, drawing level lines through these points a^1, b^1, c^1, and d^1. Lift the distances d, c, b, and a on the run of the set line from the point A, and transfer them to their corresponding level lines a^1, b^1, c^1, etc., in the Half Breadth, setting off to the left of the knuckle perpendicular. Next produce in the Body the line of the T and I frames indefinitely above the rail, also produce T and I in the Sheer until they cut the set line A B, and level over into the Body the cutting points e to T and f to I frame produced. Measure in the Body the distance of e^1 and f^1 from the centre line, and set them off parallel lines to the centre in the Half Breadth; then lift e and f on the run of the set line in the Sheer from A, and lay them out to the left of the knuckle perpendicular on their corresponding level lines e^2 and f^2. The set line can now be drawn through these spots in the Half Breadth, which is then girthed from the centre for the position of the cants, etc., marked 1 c, 2 c, 3 c, 4 c, 5 c, B, e^2 and f^2, and laid-off on the expansion set line from X. The expanded cants, which are shown in the Sheer by dotted lines E C 5, E C 4, and E C 3, as explained on page 32, have the points a, b, c, and d, levelled on to them, and are then girthed lengthways to the rail and knuckle, above and below the points, and with these girths as radii describe arcs of circles from the cant spots on the corresponding side of the expansion set line. For the position of the transom and I frames, in the expansion, girth the distances in the Body of e^1 and f^1 relative to the rail and knuckle respectively, and with these distances for radii describe arcs from the points f^2 and e^2, on the set line, for the knuckle and rail. Next girth the rail in the Half Breadth for the position of the cants, and bend the lath round the arcs in the expansion, then the points of contact will be the position of the cants, through which draw curve. Do the same with the knuckle line. The point R^1 should agree with the girthed distance on the rail of a^1 from the centre line. The cants can now be shown in the expansion through the three points. The position of the buttocks on the expansion set line are got by girthing the Half Breadth set line from the centre for their distances, and setting them off from X along X f^2. The position on the rail and knuckle edges are found by girthing in the Sheer their length to the rail and knuckle respectively above and below the set line, and describing arcs of circles from the points on the expansion set line as was done for the cants. The cutting points of these arcs on the curve will allow the buttocks to be drawn in. To prevent confusion they are left out of the Sheer. If a batten is bent round either the rail or knuckle of the Half Breadth, and the points

Fig. 46.

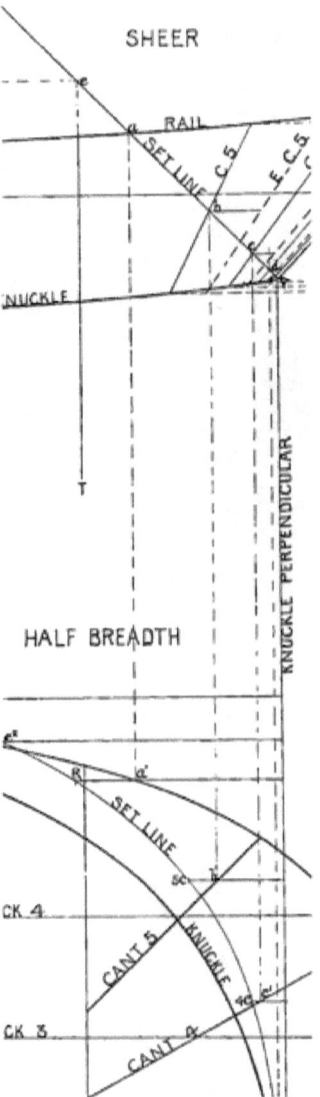

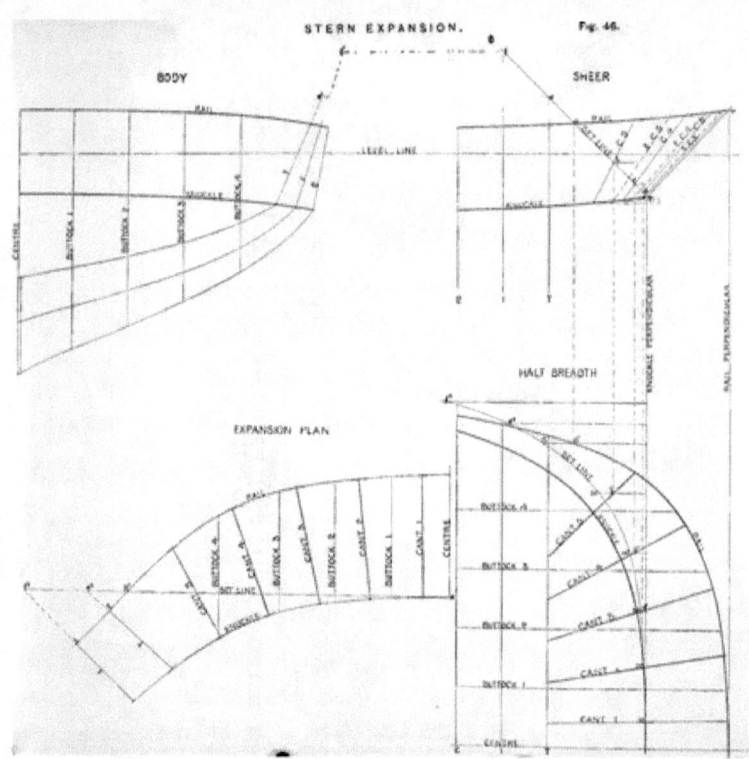

BODY

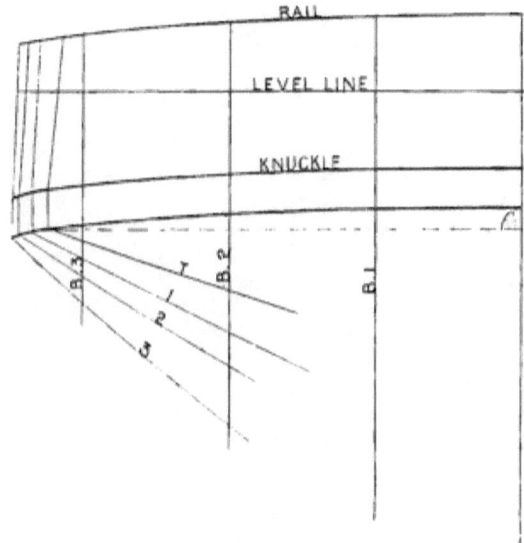

BODY

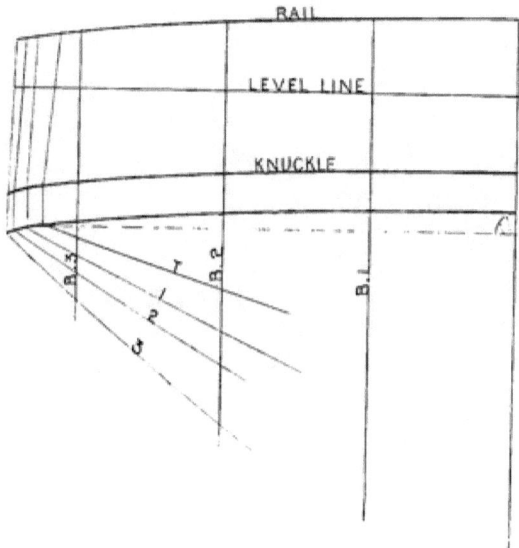

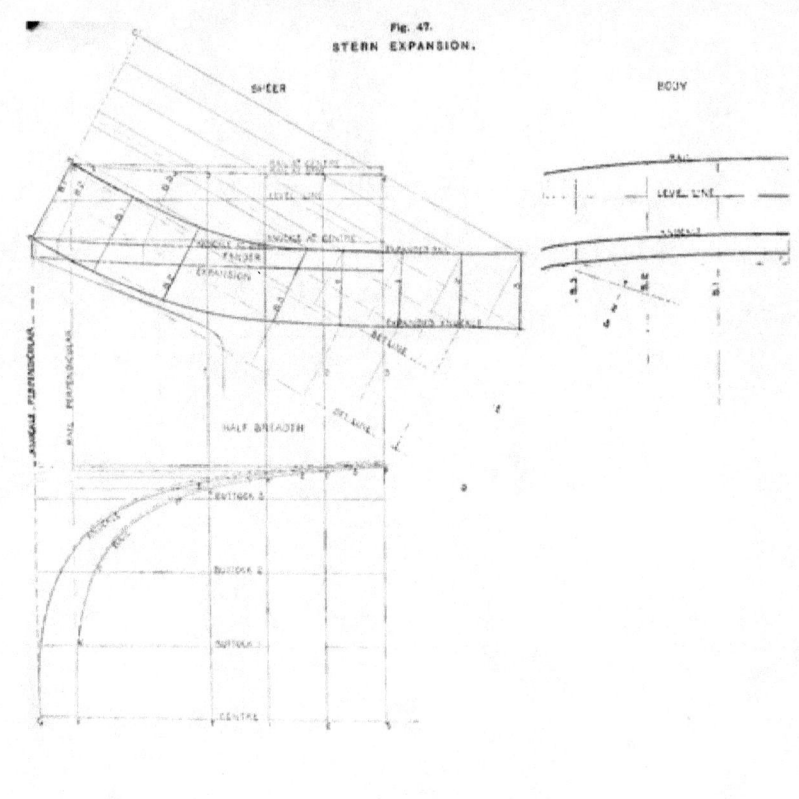

Fig. 47.
STERN EXPANSION.

of the cants and buttocks lifted and girthed round the expanded edge they should agree with the positions already indicated when the plan is correct. The butts of the plating are now arranged clear of the cants and the ordinary frames, and a skeleton template made showing the expanded top and bottom edges with cross pieces nailed on for the heel of cants, T, and I frames, with the set line marked across them. The plater extends indefinitely on the boards the line f^2 X in the direction of X, and turns the template over to find the position of R^1 on the line for the reverse side of the vessel. Allowance in ordering and working the plates should be made on the top and bottom edges for connections. The plates are ordered to embrace the curve, and care should be exercised to see that the plate is sufficiently wide to make the connections at top and bottom.

Expansion of a Tumble-home Stern.—The stern in small coasters, tug-boats, and river craft is sometimes formed as shown in **Fig. 47**. In that case the expansion is somewhat different to that already described. Show in the Sheer, Half Breadth, and Body, the initial faired-up lines, i.e., the frames, buttocks, rail, and knuckle. Join F to B and G to A. Extent indefinitely A B to C. Erect perpendiculars or set lines A D and B E of indefinite length. Through points a, b, c, d, e, f, and g drop perpendiculars on to B C, and draw level lines through the same points in the Half Breadth, a^1, b^1, c^1, d^1, etc., to the rail vertical. Lift in the Sheer the square distance of the points a, b, c, d, e, f, and g from B C, and lay them off forward of F B in the Half Breadth on their respective level lines, and trace the dotted line through the spots, which is called the corrected rail line. Girth this line from the centre for the new position of the buttocks and frames, and measure the distances out from B on B E, and from the points erect perpendiculars to cut the lines, produced, already shown through a, b, c, d, etc., which are parallel to B E. The intersection of corresponding lines will give points for the expanded rail line. Repeat this process with the knuckle, with this exception, that the corrected knuckle line in the Half Breadth is got by setting forward of G A, the distances lifted from the Sheer. The level lines, etc., are dotted in.

Join together the frame and buttock points on the expanded rail and knuckle. The position of the cant frames may be found by placing them in the Half Breadth and girthing the original or initial rail and knuckle lines for the same. They are then set along the expanded lines. The butts of the plating are fixed to clear cants. Allowance should be made for turning down on to the fender.

of the cants and buttocks lifted and girthed round the expanded edge they should agree with the positions already indicated when the plan is correct. The butts of the plating are now arranged clear of the cants and the ordinary frames, and a skeleton template made showing the expanded top and bottom edges with cross pieces nailed on for the heel of cants, T, and I frames, with the set line marked across them. The plater extends indefinitely on the boards the line f^2 X in the direction of X, and turns the template over to find the position of R^1 on the line for the reverse side of the vessel. Allowance in ordering and working the plates should be made on the top and bottom edges for connections. The plates are ordered to embrace the curve, and care should be exercised to see that the plate is sufficiently wide to make the connections at top and bottom.

Expansion of a Tumble-home Stern.—The stern in small coasters, tug-boats, and river craft is sometimes formed as shown in **Fig. 47.** In that case the expansion is somewhat different to that already described. Show in the Sheer, Half Breadth, and Body, the initial faired-up lines, i.e., the frames, buttocks, rail, and knuckle. Join F to B and G to A. Extent indefinitely A B to C. Erect perpendiculars or set lines A D and B E of indefinite length. Through points a, b, c, d, e, f, and g drop perpendiculars on to B C, and draw level lines through the same points in the Half Breadth, a^1, b^1, c^1, d^1, etc., to the rail vertical. Lift in the Sheer the square distance of the points a, b, c, d, e, f, and g from B C, and lay them off forward of F B in the Half Breadth on their respective level lines, and trace the dotted line through the spots, which is called the corrected rail line. Girth this line from the centre for the new position of the buttocks and frames, and measure the distances out from B on B E, and from the points erect perpendiculars to cut the lines, produced, already shown through a, b, c, d, etc., which are parallel to B E. The intersection of corresponding lines will give points for the expanded rail line. Repeat this process with the knuckle, with this exception, that the corrected knuckle line in the Half Breadth is got by setting forward of G A, the distances lifted from the Sheer. The level lines, etc., are dotted in.

Join together the frame and buttock points on the expanded rail and knuckle. The position of the cant frames may be found by placing them in the Half Breadth and girthing the original or initial rail and knuckle lines for the same. They are then set along the expanded lines. The butts of the plating are fixed to clear cants. Allowance should be made for turning down on to the fender.

CHAPTER VI.

Scrieve Board: Information placed upon it—Its Purpose—How Prepared—Scrieving in the Frames—Decks—Shell Plating Sight Edges—Shell Plating Inner Edges—Ribbands, Keelsons, Floors, Cant Knees—Lifting Beams—Frame Bevels—Applying Bevels—Checking Bevels—Handy Bevelling Machine—Machine Bevelling.

SCRIEVE BOARD.

The Scrieve Board is a platform formed of well-seasoned deals, laid edge to edge, fastened securely together, and placed in position near the frame furnace. It is planed on the top side, and then coated with a mixture of lamp black and liquid turpentine. When dry the Body plan is copied upon it, and the lines cut in with a "scrieve-knife." Sometimes both sides of the ship are scrieved in, with the base line of each body on opposite edges, so that the frames lap unto each other, but to prevent confusion it is better to place each full Body on separate boards, or only to scrieve half of the ship, as shown in **Fig. 48.**

The Information placed upon it.—The moulded shape of every frame, floor, and tank knee. The position on the frames of all decks, platforms, shell plate edges, keelsons, ribbands, harpins, side stringers and girders, keels, stern cants, screw boss, number of each frame, and whatever comes in contact with the frames, reverses, or floors.

Its purpose is that every frame, reverse, floor, and beam may be turned by the platform men to their proper shape, and punched and marked for receiving the other parts of the structure.

How Prepared.—Strike in with a chalk line the base of the after body, and erect with trammels a perpendicular to it for the centre line on, or about, the middle of the board. Then place in position lines for the half breadth, depth moulded, rise of bottom, half siding of stern post, level or water-lines, buttocks, and diagonals. Transfer on a short batten from the Sheer on the loft floor heights of the rail, knuckle and deck at side above the depth moulded line at every frame, and strike in level lines through the spots; the after sheers over the after body, and the forward sheers over the forward body. Then lift from the Half Breadth the half beam on every frame at rail, knuckle and deck, and place them on their corresponding sheer heights just lined in. Do the same with the $\frac{1}{2}$ ordinates on the level lines and on

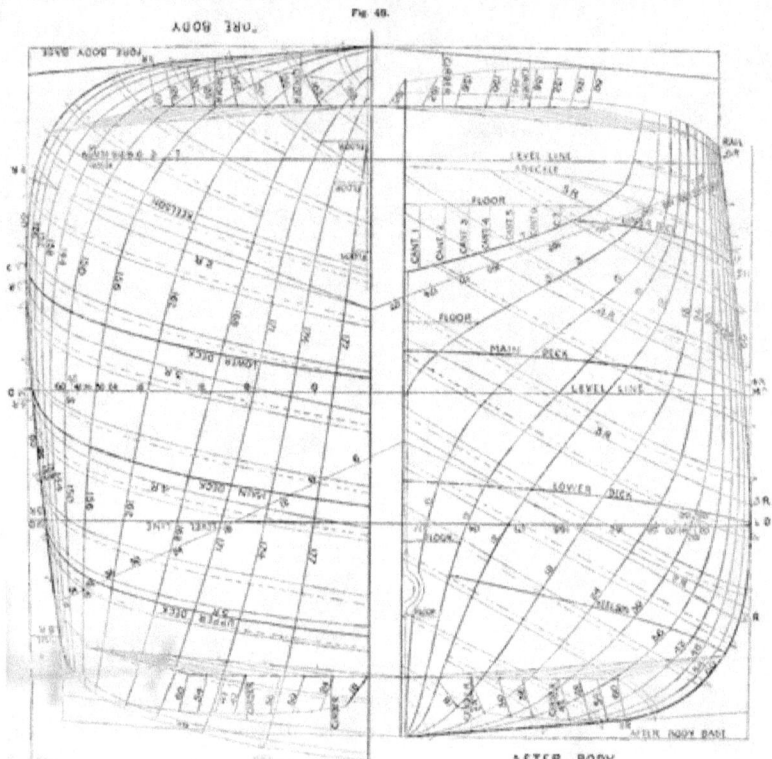

Fig. 48.

the diagonals; also lift from the Sheer above the base the points of intersection of the frames and buttocks, and the height of the after frame feet on the inside of the stern post, and transfer them unto the board on the buttocks and the half siding of the post respectively. Where the spots come very close, it is usual only to lift every fifth frame and divide the others in suitably by inspection. The points are marked on the scrieve board with a sharp-pointed loft nail, and afterwards chalked in and numbered for clearness.

Scrieving in the Frames.—Pin a suitable batten—thick at the ends and thin about the middle for the bilge curve—to the midship frame spots; fair it carefully and scrieve in. With this as a guide, in conjunction with spots, French chalk in fair for about quarter length amidships every fifth frame. The intermediates being close are divided in on a graduated scale—this is frequently done on a short batten—then the scrieving in of every frame from the midships is proceeded with. When the after body is scrieved, a base line for the fore Body is lined in on the opposite edge of the board, in such a position that some of the level lines of the after Body extended across will work in. The centre line is kept away from that of the after Body which allows the reversing spots of the after Body to be marked clearly over the fore Body as shown. The initial lines, as in the after Body, are placed in position, and the midship section copied from the after Body by making a wood template of the form. The spots for the water-lines, diagonals, bow lines, sheer heights, and frame feet are lifted from the loft floor, and set off on the scrieve board, and the process carried out as in the after Body.

The frames are usually first scrieved in without regard to the position of the screw boss or bosses, or any projection, and the alteration set forth on page 12 for the boss carried out.

Many of the shipbuilders lay off their vessels, in the office, on $\frac{3}{4}$ inch scale, on the diagonal system, and supply the loftsman with ordinates on these diagonals, which are shown in **Fig. 49**. In that case the positions of the diagonals up the centre, on the base, and on the half moulded breadth line are given the loft with the ordinates on the run of the diagonals, instead of the level line half ordinates.

In other yards the ship is laid down on $\frac{3}{4}$ inch scale, in the office, as described in Chapter I., and ordinates and particulars supplied loftsman in a book from this plan for "scrieving ship in." The after end, for a distance of about 20 feet forward of the transom, being faired-up full size on the loft floor at the same time as the stern. It is not our

the diagonals; also lift from the Sheer above the base the points of intersection of the frames and buttocks, and the height of the after frame feet on the inside of the stern post, and transfer them unto the board on the buttocks and the half siding of the post respectively. Where the spots come very close, it is usual only to lift every fifth frame and divide the others in suitably by inspection. The points are marked on the scrieve board with a sharp-pointed loft nail, and afterwards chalked in and numbered for clearness.

Scrieving in the Frames.—Pin a suitable batten—thick at the ends and thin about the middle for the bilge curve—to the midship frame spots; fair it carefully and scrieve in. With this as a guide, in conjunction with spots, French chalk in fair for about quarter length amidships every fifth frame. The intermediates being close are divided in on a graduated scale—this is frequently done on a short batten—then the scrieving in of every frame from the midships is proceeded with. When the after body is scrieved, a base line for the fore Body is lined in on the opposite edge of the board, in such a position that some of the level lines of the after Body extended across will work in. The centre line is kept away from that of the after Body which allows the reversing spots of the after Body to be marked clearly over the fore Body as shown. The initial lines, as in the after Body, are placed in position, and the midship section copied from the after Body by making a wood template of the form. The spots for the water-lines, diagonals, bow lines, sheer heights, and frame feet are lifted from the loft floor, and set off on the scrieve board, and the process carried out as in the after Body.

The frames are usually first scrieved in without regard to the position of the screw boss or bosses, or any projection, and the alteration set forth on page 12 for the boss carried out.

Many of the shipbuilders lay off their vessels, in the office, on $\frac{3}{4}$ inch scale, on the diagonal system, and supply the loftsman with ordinates on these diagonals, which are shown in **Fig. 49**. In that case the positions of the diagonals up the centre, on the base, and on the half moulded breadth line are given the loft with the ordinates on the run of the diagonals, instead of the level line half ordinates.

In other yards the ship is laid down on $\frac{3}{4}$ inch scale, in the office, as described in Chapter I., and ordinates and particulars supplied loftsman in a book from this plan for "scrieving ship in." The after end, for a distance of about 20 feet forward of the transom, being faired-up full size on the loft floor at the same time as the stern. It is not our desire to express any opinion respecting the merits of the various systems; they have been found to answer the purpose effectually, or they would not be continued.

Fig. 49.

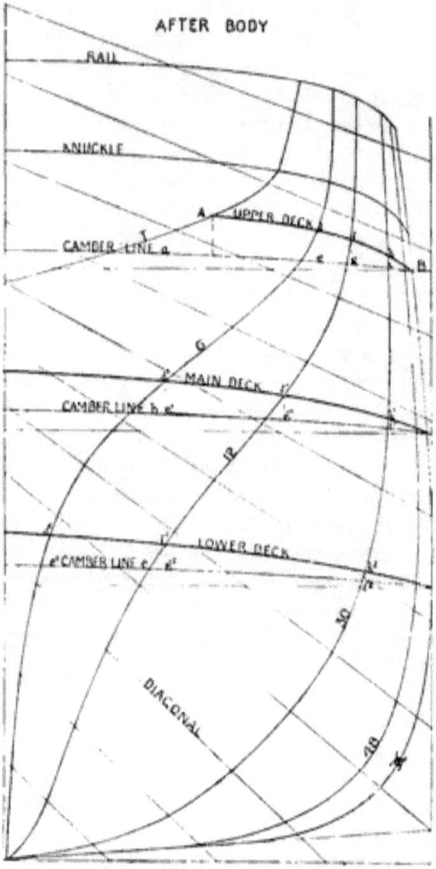

AFTER BODY

Scrieving in the Deck Lines.—The upper deck sheer at side above the depth moulded line, on about every fifth frame, is lifted from the loft floor on a batten, and transferred on to the respective frames on the scrieve board, and the line scrieved in. The main and lower deck side lines are, sometimes, made parallel to this line, which means that the centre 'tween deck height is a variable quantity and open to objection. A better plan is to make all the decks with camber parallel to one another at all points. This may be done by placing the upper edge of the beam camber mould fair with the midship side line, and its centre on that of the Body; and setting the centre of a duplicate mould fair with that of the ship and its top edge parallel to the top edge of the first mould and at a suitable distance down, i.e., height 'tween decks; mark the point where the second mould cuts the midship frame. Then lift on a batten the perpendicular heights from the upper mould edge to the deck line, and set them off, on their respective frames, parallel to the centre above the second mould edge, which will give points for the main deck line. Repeat the process for the lower and other decks, if cambered.

Fig. 49 shows, perhaps, an easier method of doing this. Let A B be the side line of the upper deck. Then set off camber lines—due to width of the ship at the deck—a, b, c, on the midship frame at each deck, which may be done with the beam mould. Lift on a batten, for frames 6, 12, and 30, $d\,e, f\,g, h\,i$ above camber line a of the upper deck, and place the distances on their corresponding frames above the camber line b of the main deck, $d^1 e^1, f^1 g^1, h^1 i^1$ respectively. Draw curve through the spots and you have main deck side-line, which fair-up and scrieve in. Repeat the process at the lower deck if necessary. The fore body deck lines are formed in the same way.

Scrieving in the Shell Plating Sight Edges.—The width of each strake on the midship frame, with the breadth of the edge laps, and the distance above the deck of the sheer strakes, together with a tracing showing the run of the plate edges is usually given to the loftsman. The midship widths he places around the midship section as near as possible. The sheer and garboard strakes, and the flat keel plate admit of no alteration and are carried, if possible, forward and aft about the same width. Any addition or lessening of the widths must be made on the thin strakes. To get the curve of the sight edges for the scrieve board, straight lines $a\,b, c\,d, e\,f, g\,h$, in **Fig. 40**, are drawn across the $\frac{1}{4}$ inch scale body sections from the plate sight edge on the midship section to the centre line; and the distance of the plate edge of each strake is girthed on the sections from this line, and on the copy given to the loftsman these dimensions and lines are marked; so that it is an easy matter to mark off the lines on the scrieve board and set therefrom the position of the plate edges.

Amidships, the points may require a little manipulation to suit the correct sheer, which, in most cases, will not agree exactly with the ¼ inch scale Sheer Draught. After this is done, the lines are faired-up and scrieved in. Care should be taken that the (girthed) width of the boss or furnace plate is not too wide for the furnace. Many of the shipbuilders give the loftsman the girthed widths of the sight edges on about every fifth section, with heights on the posts, which are set round the scrieve board frames, and after fairing are scrieved in.

Scrieving in the Extreme Edges of the Inside Strakes.—This is easily done by girthing round the frames from the sight edges the width of the lap and scrieving a line through the spots in each case. The lines are shown dotted in **Fig. 48** to distinguish them ; the usual practice is to paint the lap in.

Scrieving in the Ribbands.—They are arranged, as in **Fig. 48**, on the outside strakes clear of the decks and inside strakes, if possible, for the entire length—marked 1 R, 2 R, 3 R, etc.

Scrieving in the Side Keelsons.—In a ship with turned-up floors the curved line will commence from the top edge of the floor where the intercostal cuts. Those clear of a double bottom from the moulded edge of the frame ; except in the case of a keelson not extending to the shell plating, then the width of the frame must be set in for the starting point of the curved line.

Scrieving in the Cant Frame Knees.—The moulded edge of the vertical bar attaching knees to the transom floor is scrieved in, shown by marks 1 C, 2 C, 3 C, etc., in **Fig. 48**. The positions are lifted on the run of the transom from the Half Breadth on the loft floor.[*]

It remains to add that whatever has to be attached to the frames, reverses, or floors, must be distinctly shown on the scrieve board. This will decide any doubtful point.

In the sketch, **Fig. 48**, to make it clear, only occasional frames are shown, but every frame, floor, etc., must be scrieved-in in an actual case.

Scrieving in the Floors.—In a ship with *turned up floors* they are simply copied from the Body sections, which is fully explained on page 37, the intermediates being either divided in or lifted on each buttock from the faired longitudinal lines and scrieved in. In cases where the floors are ordered in a hurry from a small scale plan

---

[*] In battle-ships the side armour, and backing lines, together with longitudinal bulkheads, are scrieved in on the boards.

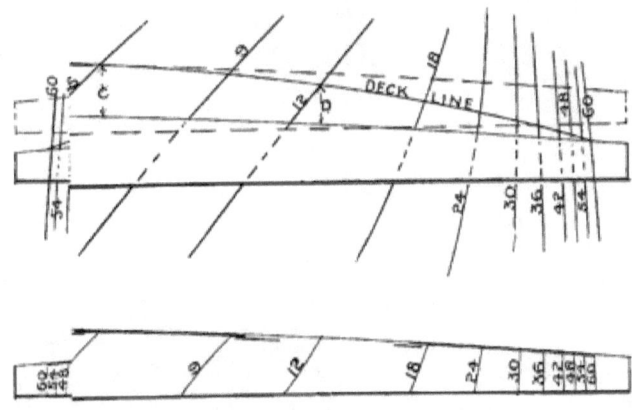

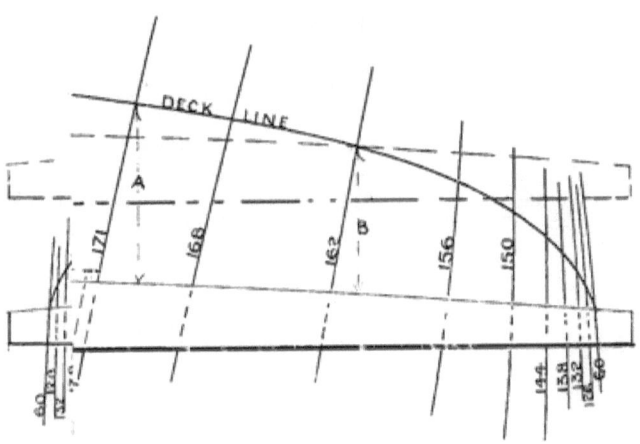

Fig. 50.
AFTER BODY

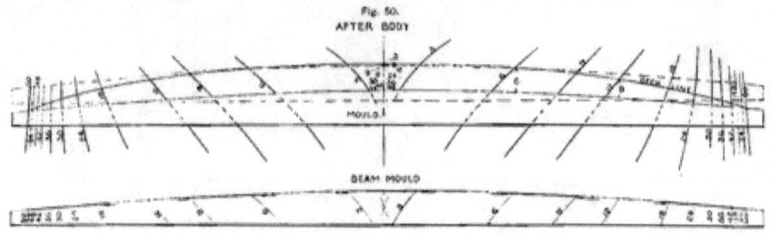

BEAM MOULD

FORE BODY

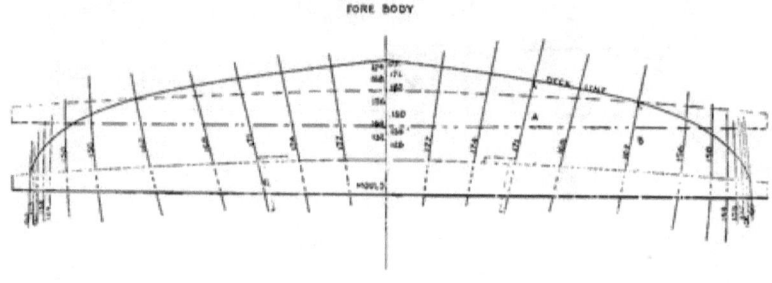

BEAM MOULD

before the ship is laid off properly (or very much decided), it is best to carry out on the scrieve board the method described on page 38, and check the ordered sizes of the plates to prevent unnecessary delay. The loftsman, therefore, is supplied with a copy of the ordered floor specification. When a *cellular double bottom* is fitted in a merchant ship the floor top is usually level right fore and aft. In flat bottomed ships it may be cambered slightly to allow the floors to be ordered as sketches under 9 inches taper. The inner bottom being flat it is chalked in parallel to the base line, and on it is set off for each frame the knuckle points. The cutting line on the frames is transferred on to the boards, and a curve passed through the spots. The knuckle and frame points of the side are connected, and care taken that these lines are square to the frame surface in each case. When chalked in they are faired on the loft floor in the manner described on page 41, and then scrieved. The knees outside of the double bottom are next placed in position. For this purpose strike in the "diminishing line," as explained on page 44, on which all the knees terminate at the frame; and join knuckle point of the tank side with corresponding frame on this diminishing line.* The side girders in the double bottom are placed in position clear of the shell plate edges and made parallel to the centre line in merchant ships. *Extra* girders are usually fitted in the engine space. The entire double bottom is then scrieved. For the purposes of testing frames, etc., their position on level and diagonal lines are shown on the opposite side, together with the tank side and knees (see **Fig. 48**). In the case of the cellular *double bottom* of a *war vessel*, great care is taken in fairing-up the inner bottom and longitudinals on the loft floor, so that it only becomes necessary to copy the body plan on to the scrieve board. The system of fairing is described on page 130.

Lifting the Length of the Beams.—The loftsman is supplied with a list of the beam frames on each deck. These he marks distinctly with chalk on the scrieve board for a short distance below and above the deck line. One mould may be made to serve the purpose of each deck, if the different bodies are put on reverse sides of the mould. The method of lifting is to place the mould centre in **Fig. 50** fair with the centre line of the scrieve board and touching the deck line, at each side, on the midship frame. Then lift on a lath square up from the edge of the mould the distance of the deck line on the beam frames, like A, B, C, and D, and set these distances off

---

* The height of knee at the knuckle may be slightly below the inner bottom.

before the ship is laid off properly (or very much decided), it is best to carry out on the scrieve board the method described on page 38, and check the ordered sizes of the plates to prevent unnecessary delay. The loftsman, therefore, is supplied with a copy of the ordered floor specification. When a *cellular double bottom* is fitted in a merchant ship the floor top is usually level right fore and aft. In flat bottomed ships it may be cambered slightly to allow the floors to be ordered as sketches under 9 inches taper. The inner bottom being flat it is chalked in parallel to the base line, and on it is set off for each frame the knuckle points. The cutting line on the frames is transferred on to the boards, and a curve passed through the spots. The knuckle and frame points of the side are connected, and care taken that these lines are square to the frame surface in each case. When chalked in they are faired on the loft floor in the manner described on page 41, and then scrieved. The knees outside of the double bottom are next placed in position. For this purpose strike in the "diminishing line," as explained on page 44, on which all the knees terminate at the frame; and join knuckle point of the tank side with corresponding frame on this diminishing line.* The side girders in the double bottom are placed in position clear of the shell plate edges and made parallel to the centre line in merchant ships. *Extra* girders are usually fitted in the engine space. The entire double bottom is then scrieved. For the purposes of testing frames, etc., their position on level and diagonal lines are shown on the opposite side, together with the tank side and knees (see **Fig. 48**). In the case of the cellular *double bottom* of a *war vessel*, great care is taken in fairing-up the inner bottom and longitudinals on the loft floor, so that it only becomes necessary to copy the body plan on to the scrieve board. The system of fairing is described on page 130.

Lifting the Length of the Beams.—The loftsman is supplied with a list of the beam frames on each deck. These he marks distinctly with chalk on the scrieve board for a short distance below and above the deck line. One mould may be made to serve the purpose of each deck, if the different bodies are put on reverse sides of the mould. The method of lifting is to place the mould centre in **Fig. 50** fair with the centre line of the scrieve board and touching the deck line, at each side, on the midship frame. Then lift on a lath square up from the edge of the mould the distance of the deck line on the beam frames, like A, B, C, and D, and set these distances off

---

* The height of knee at the knuckle may be slightly below the inner bottom.

on the centre line above the edge of the mould and number the points. Then scrieve the form of the midship frame, for the port and starboard sides, on the mould; setting in the line about ⅜ of an inch owing to the beam arm having to go into the bosom of the frame. There are cases where the beams are on the heel of the frames. Then shift the mould up to the next beam frame at the deck line, keeping

Fig. 51.

AFTER BODY

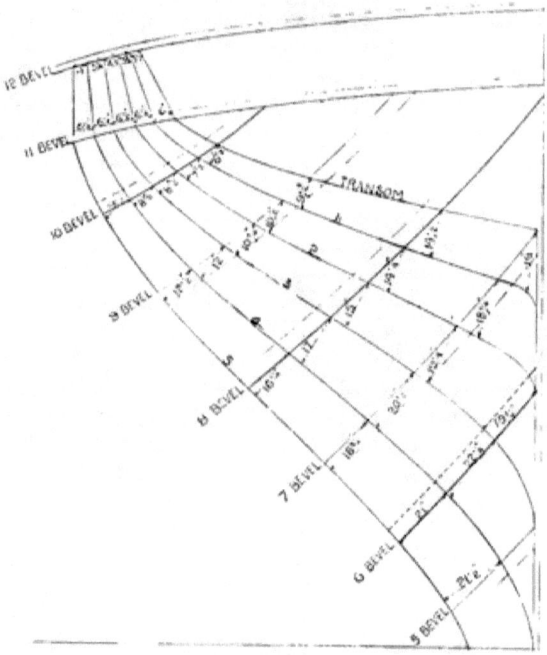

the centre right, and mark it in the same manner. This process is repeated until all the beams of each body are scrieved on the mould to the form of the frames. In **Fig. 50** the mould is shown dotted in position when lifting numbers 6 and 162. The centre heights are not necessary when both sides of the ship are scrieved, but if only half of the body is shown they are indispensable; the mould must be carefully placed on the midship frame before lifting the heights, *i.e.*, level across and centred. After scrieving one side, the scrieve is reversed with care to the other side, by bending a lath round the edge of the mould and transferring the spots. The form of the frame from these points is got by an adjustable set square marked E, and by noting the amount of round on the frame. The beams are then numbered, and the ship's number and the name of the deck scrieved in. When an iron deck has to be fitted the laps of the plating are marked on the edge

Fig. 52.

FRAME BEVELS. No. SHIP.

No. of Frame.	BEVELS IN INCHES.											
	1	2	3	4	5	6	7	8	9	10	11	12
Transom.							19	11½	9½	6½	6	3½
1							18¾	14¼	10½	7½	6¼	3¾
2						19¾	19¼	15	10¾	8½	6½	4½
3						22¼	20½	17	12	8½	6¼	5¼
4					21½	21	18¾	16¼	12½	9½	6¼	4
5												
6												
7												
8												
9												
10												
11												
12												

of the mould. In some cases for a wood deck the widths of the planks are supplied, which being marked on a lath are given to beam-turner to punch horizontal flange of the beam. Each beam is set in the squeezes to the shape of this mould.

The holes for the beam arms are marked from a duplicate template used on the boards for punching the frames, so that the beam and frame correspond.

All beams are not cambered; in the case of wide spaced hold beams they act structurely more to the purpose of strength by being straight across the ship. In the case of web frames the beam arm

Fig. 53.

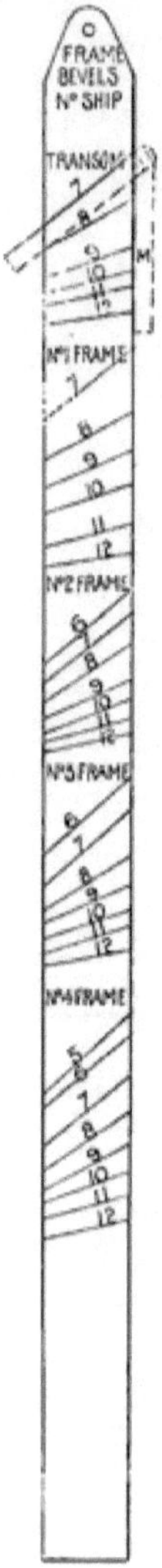

will stand in from the heel of the frame more than ¾ of an inch; in certain classes of web frame beams the beam is attached to the frame. Perhaps the wisest plan is to give a separate mould for the web frame beams.

A separate camber mould, with only the centre line marked on, is given for bulkheads. A lath, showing the length of the beams, is also sometimes given out with the moulds by the loftsman.

Lifting the Bevels of the Frames from the Scrieve Board.—Where the work of bevelling the bars is done by manual labour the bevels are taken at each plate edge, or at the ribbands. **Fig. 51** shows an enlarged sketch of a few frames in the after body. Lift the distance, as shown, between consecutive scrieves, square off from the frame you are dealing with, and enter them on a sheet of paper, or in a book (see **Fig. 52**). A few bevel boards are made of soft wood, like **Fig. 53**, 4 to 5 feet long, and of the width of the frame shell flange. In **Fig. 54** is given a sketch of a wood machine for marking the bevels on the board. The different spacing of frames, from 20 to 26 inches, is lined on the centre line of the machine, measured from A. The screw B is fixed on the centre through the groove in the movable leg M to the fore and aft spacing of the frames from heel to heel of bars. A scale of inches is marked along the bottom piece to take all possible distances between the scrieves. This scale is only on the right-hand side of the centre, because the bevel must always be standing or open. The bevel board is placed in position, as shown, behind M, and the front edge of the leg M is moved to the distance between the scrieves given in **Fig. 52**, and a line drawn across the board and numbered: this mark gives the distance the bar has to be opened from the heel to the top, or the angle of opening (see "a" in small **Fig. 55**). The process is repeated until both sides of the board are filled with bevels (given in **Fig. 53**), when other boards are secured and filled until all the frames are lifted.

SCRIEVE BOARD.

Applying the Bevels to the Frame Bars.—Lift the bevel from the board by an adjustable set square in the manner shown in the **Fig. 53**, and apply it when bevelling square to the *heel* or back of the bar when it is lying on the bending slab (as shown in **Fig. 55**) with the arm B laid on the slab.

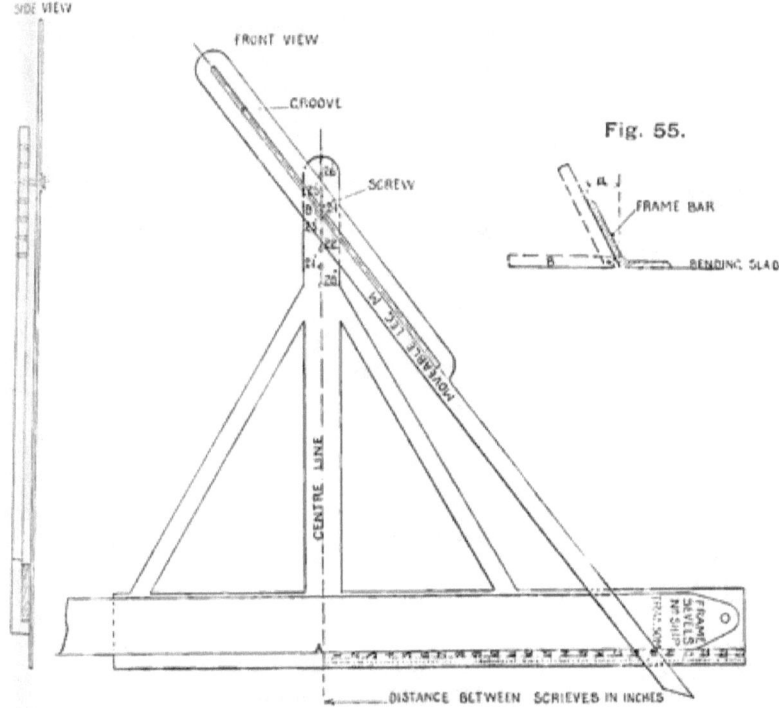

Fig. 54.

Fig. 55.

Checking the Frame Bar Bevels on the Boards.—A light iron skeleton machine (**Fig. 56**), sometimes called the "detective," is used for this purpose. The frame bar is placed on the scrieve line and the machine, fixed to the correct spacing of the frames, is made to stand *square off* to the bar, with the point A on the next scrieve. Then the front edge of the movable leg, with its point at the heel of the frame, should touch for the full depth of the flange if the bevel is correct. All frames are tested in this way before erection.

64　　　　　　　　NAVAL ARCHITECTURE.

Handy Bevelling Machine.—Fig. 57 shows another bevelling machine made, to ¼ full size, of hard wood. A narrow brass plate A is fixed along the edge in which holes are bored and tapped for fixing the

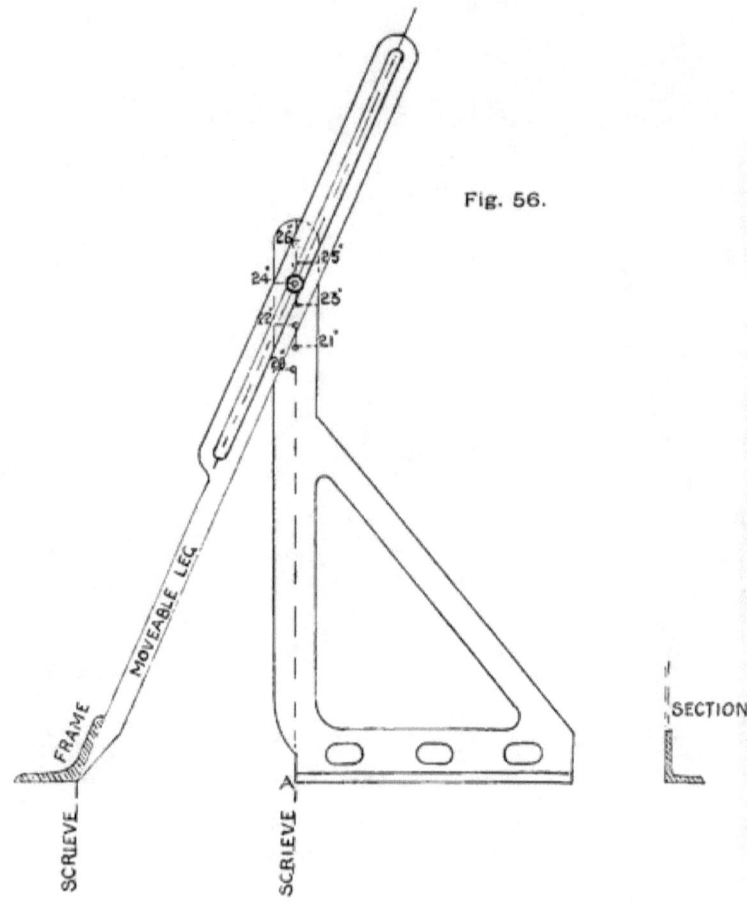

Fig. 56.

screw D to the correct spacing of the frames. On the edge B C is set off inches and parts of an inch. It is used in the same manner as the one already described.

Machine Bevelling.—Most of the large shipbuilding establishments have bevelling machines, standing in front of the furnace, for bevelling the bar as it is drawn out of the furnace. The bevels for this purpose must be in degrees of a circle. These are got by applying

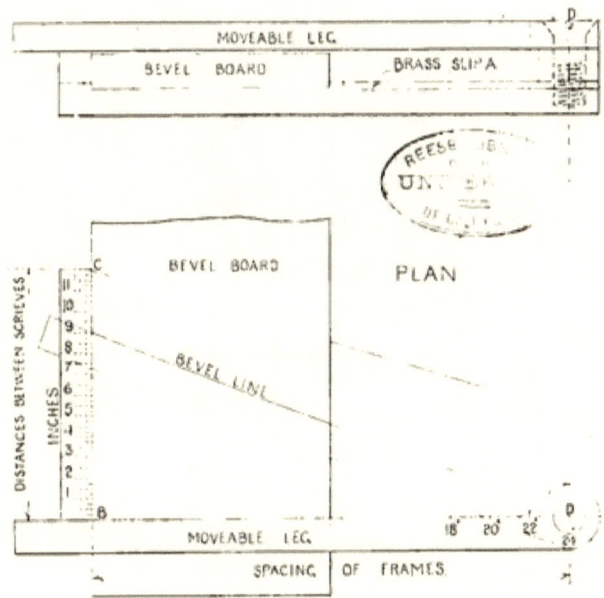

Fig. 57.
END VIEW

to the lines on the bevelling board a quadrant with a projecting ledge, and obtaining the degrees you mark them with chalk on the dial placed on the front of the bevelling machine. Fig. 58 shows the quadrant set on the board to bevel 7, which is 20½ degrees. It is only necessary to say that the bevels are taken at uniform distances apart, usually 4 to 5 feet, beginning at the deck or centre, whichever is most convenient.

Fig. 58.

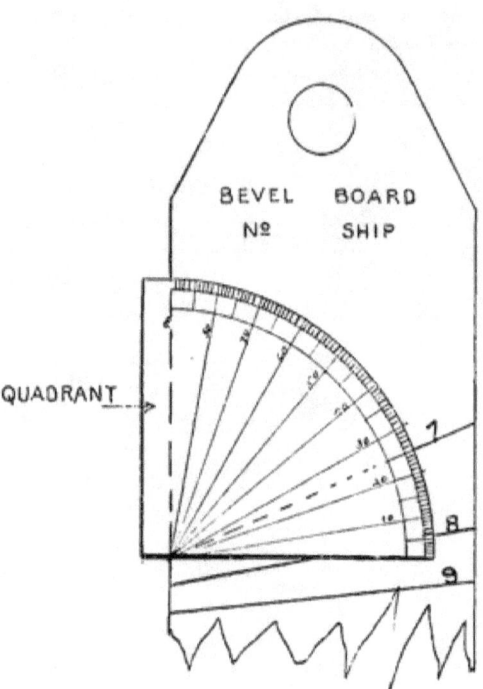

CHAPTER VII.

Ribbands; Form of a Ribband Line—Stem Termination—Stern Termination—Laying Ribband Lines Off and Marking Battens—Deck and Inner Bottom Ribbands. *The Common Harpin*: Form of the Moulded Edge—Form of the Bevelled Edge—Bevelling Board. *The Sheer Harpin*: Form of the Moulded Edge—Form of the Bevelled Edge—Bevelling Board—Expansion of Moulded Edge in the Sheer. *The Stern Harpin*: Form of the Moulded Edge—Form of the Bevelled Edge.

RIBBANDS AND HARPINS.

For erecting purposes, to keep the frames in their relative positions until the inside strakes of the shell-plating are in place and secured, ribbands and harpins are used. The former, which are made of pine wood about 6 inches siding, are placed over the entire length of the ship on the outside of the frames, in way of the outside strakes of the shell-plating. They are usually straight lines in the after Body sections, and curved to the sheer in the fore Body for about the half length, then run in straight to the stem. The line placed on the scrieve board is the top edge of the ribband. Where the vessel is bluff at the ends, and it would be difficult to bend the ribband to the form of the frames, harpins made of angle bar are turned to the *expanded* shape. There are three kinds of harpins: first, the common harpin, which appears straight in the Body plan, the same as a ribband: second, the sheer harpin, which takes the sheer of the top sides, and, of course, appears curved in all the plans: third, the stern harpin, which is level in the Body and Sheer plans.

The True Form of a Ribband Line.—It is found by bending a light lath round the ribband line on the scrieve board, or the Body on the loft floor, and girthing relative to the centre line, the position of the half siding of the stem and stern posts, and the first ten frames at the ends; afterwards, about every fifth will answer the purpose. Lay these distances off on the loft floor, in the manner to be hereafter described.

To Find the Termination on the Stem.—Lift the position of the ribbands on the scrieve board from the base line on the stem half-siding line, and set the heights square up on the inside of the stem in the Sheer on the loft floor. Square these points down on to the Half

Breadth centre line, and place on these perpendiculars the girthed board width from the centre to the half-siding in each case, which give the terminations of the ribbands at the fore end.

To Find the Termination Aft.—When the ribband lines are below the foot of the transom, the same process described for forward is adopted; but should the ribband cut the transom frame, then lift the distance of the point square out from the centre line, and lay it out on the same frame in the Half Breadth on the loft floor. This is the termination on the transom; the other points will be got by marking on a lath this distance and -swinging the lath round—keep this point fair on the transom of the scrieve board, and proceed to girth the position of the frames forward on the run of the line in the usual way.

Laying the Ribband Lines Off and Marking the Ribband Battens.—The girthed distances, lifted from the scrieve board, are set off in the Half Breadth on the loft floor on their respective frames, and a ribband batten, about 40 feet long by $1\frac{3}{4}$ inches square, made of soft pine, is bent round the spots, and when secured, the position and number of every frame is marked upon it with a joiner's soft pencil. Care should be taken to get the marks square to the edge of the batten, so that either edge can be used for marking off. When it is filled on one side, turn it over to another, and mark in like manner by a continuation of the ribband line, until the relative position of the frames on the run of the line right fore and aft are lifted. One batten is usually sufficient for a ribband, which has its number, and the ship's number, marked upon it. The process is repeated for each ribband line; and when the entire lot is completed, they are given to the shipwrights, who mark on the top side of the ribbands the position of each frame. Sometimes they are scrieved in. Holes are bored in the ribbands for bolt attachment to the frames, and they are hoisted into the position *marked* on the frames by the frame-turner (the two marks being brought together), and well secured: where they remain to keep the ship in shape until the inside shell-plating is attached.

N.B.—The above is the ordinary way of lifting the position of the ribbands; but it is more correct at the ends, when the form of the ribband is placed on the loft floor, to set out from the line half the width of the ribband, and bend the ribband battens round these spots for the true length and position of the frames.

Deck and Inner Bottom Ribbands.—Ribbands are also fitted on the upper decks and inner bottom to keep the decks and bottom in form until part of the plating is attached. Those for the tank sides may be lifted from the loft floor, but it is not usually done. It may be noted here that the bevels for the tank side bars may be lifted from the scrieve board in the ordinary way; which does away with the practice of bending a ribband round tank side in the ship.

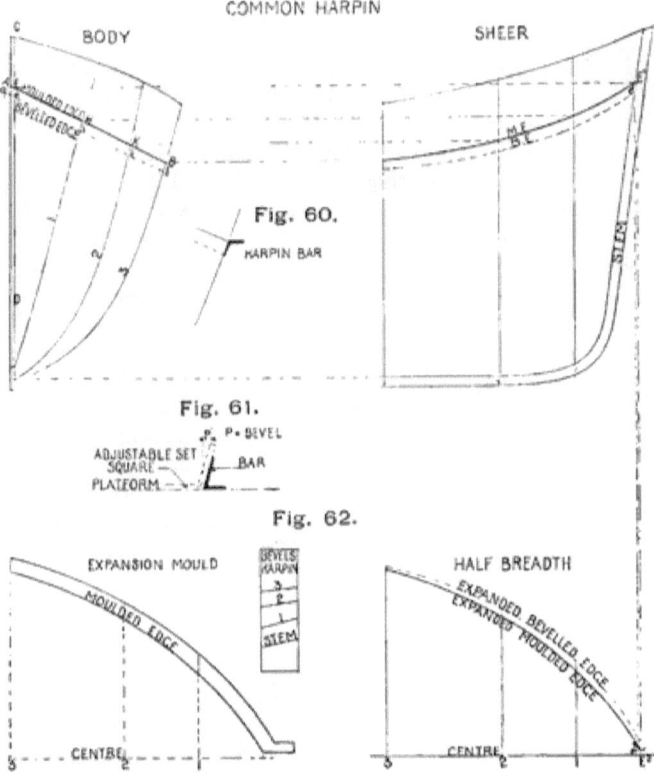

Fig. 59.

COMMON HARPIN

Fig. 60.

Fig. 61.

Fig. 62.

The Common Harpin is a continuation of an ordinary ribband, its purpose and form being the same; only it is made of angle bar.

The True Form of the Moulded Edge.—Referring to Fig. 59 place the moulded edge, A B, of the harpin in the Body, and show in the half siding of the stem C D, and level its intersection with the harpin at E into the Sheer, cutting inside of the stem at E^1. Square this point down into the Half Breadth, E^2. Lay a batten on A B, and lift from the centre line at A the distances A E, A H, A K, A B, and place these in the Half Breadth, A E on E^2 E^1, and the others on their respective stations 1, 2, and 3. Draw a curve through the spots which gives the expanded moulded edge. A pine mould is made of this line having marked on it the position of the frames and stem termination, for the use of the platform men in bending the bar to its true form. The height above the base on the stem is also figured upon it.

The Form of the Bevelled Edge.—Set down below the moulded edge in the Body on the run of the frames the depth of the vertical flange of the harpin bar E e, H h, K k, B b. Pass a curve through the spots. Draw A a perpendicular to A B, and level over the point e into the Sheer, e^1, on the inside of the stem, and drop a perpendicular from the point on to the Half Breadth centre. Lay a batten on a b, and lift from a the distances e, h, k, and b, and set them out from the centre Half Breadth line, a e on e^2 e^1 and the other distances on their respective stations. Dot the line in, for distinction, which gives the form of the bevelled edge. The difference between this and the moulded line taken at any point square to the moulded edge is the bevel required for the vertical flange, the flange standing out from the ship's side being level (see **Fig. 60**).

The Bevelling Board.—**Fig. 62.** A piece of board the width of the vertical flange is supplied with the mould showing the bevels at each frame and at the stem point. If the bevelled edge is inside of the moulded edge in the Half Breadth, then the bar must be "open bevel," if outside shut or "closed bevel." The bevel is lifted from the board and applied to the heel of the bar by a small adjustable set square. Holes are punched in the vertical flange at each frame for attachment. **Fig. 61** shows the manner of applying the bevels, which are lifted from the bevel board the reverse way to that of marking them on it. The amount of bevel applied to the bar being the distance P plus a right angle, except when less than a right angle.

The Sheer Harpin, which is also made of angle bar, may be fitted at the fore end of the vessel, being made *parallel* to the ordinary deck line. Draw in the Body the moulded and bevelled edges, and project them into the Sheer and Half Breadth plans. In **Fig. 63**, A E is the moulded edge in the Sheer. The moulded edge is shown a deep black line, and the bevelled edge a fine line.

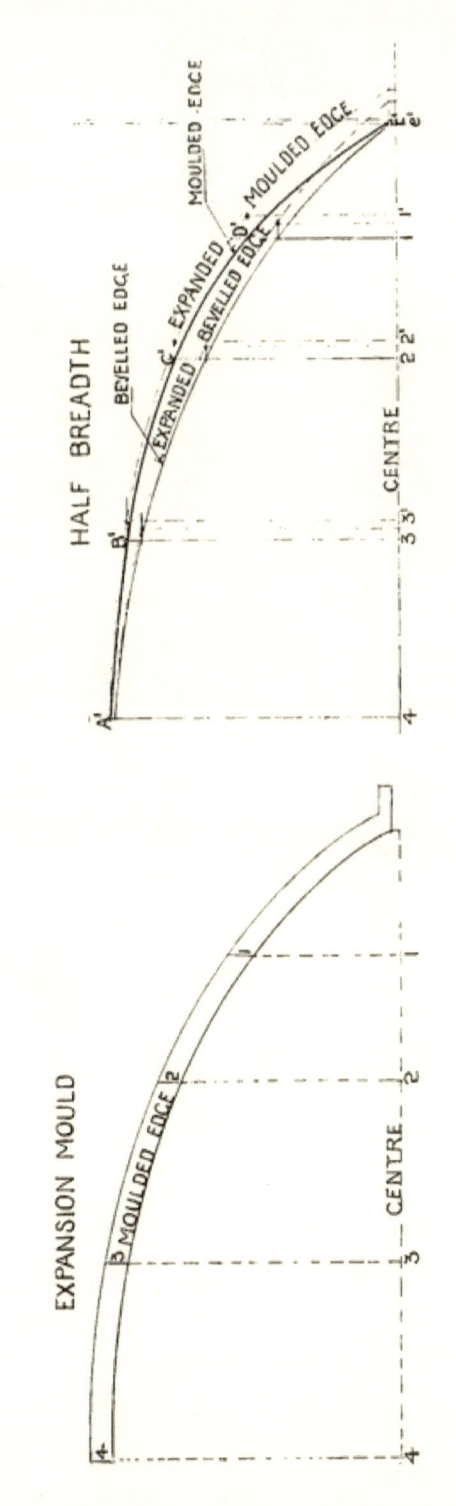

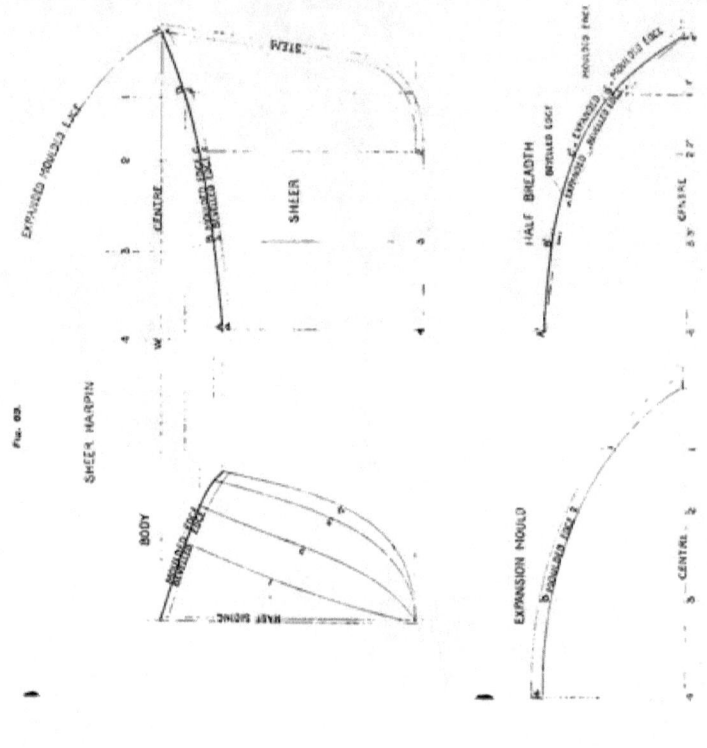

The True Form of the Moulded Edge.—Square down on to the Half Breadth centre line the point E, where the harpin cuts the inside of the stem. Set on $E^1 E$ the half siding of the stem at E, which gives the terminating point. Next girth from the point E, in the Sheer, the positions of D, C, B, and A. Lay these distances off from number 4 frame along the centre line of the Half Breadth, and draw up perpendiculars 3^1, 2^1, 1^1, and produce level lines from B^1, C^1, D^1, E^1, the cutting points of the original frames with the harpin in its true position, until they meet the new perpendiculars, which are the true positions of the frames in the expanded form. A curve passed through these intersections is the expanded moulded edge. A yellow pine mould is made to this line, and the expanded position, number of the frames and ship marked upon it.

The Bevelled Edge.—Square down into the Half Breadth the point e, and on $e^1 e$ set off from the centre the half siding of the stem at e, which gives the terminating point. Lay a batten along the bevelled edge in the Sheer and mark on it the positions of a, b, c, d, e. Place the batten along the centre of the Half Breadth, keeping point a fair with 4 frame, and mark corrected position of the frames. Draw up perpendiculars from these points to cut level lines produced from the intersection of the frames with the bevelled edge in its true position. A curve through the spots will give the expanded bevelled edge, and the difference between this line and the moulded edge at any point, taken square to the latter is the amount of bevel placed on the bar.

The Bevelling Board.—Fig. 64. A piece of pine the width of the vertical flange of the harpin, having on the bevels taken at each frame is supplied with the mould. Holes are required to be punched in the vertical flange for temporary attachment to the frames. The position of the harpin is also marked on the frames and stem when they are turned ; so that there is no difficulty in placing the harpin in its right position on the ship.

Fig. 64.

BEVELS HARPIN
4
3
2
1
STEM

Expansion of the Moulded Edge in the Sheer.—Girth the position of D, C, B, and A from E on the Sheer line E A, and set them along level line E W from E. Produce perpendiculars from the points, which give expanded position of the frames 1, 2, 3, and 4, on which set the level distances of the frames and half siding of the stem lifted from the Body. A line run through the spots gives expanded moulded edge.

The True Form of the Moulded Edge.—Square down on to the Half Breadth centre line the point E, where the harpin cuts the inside of the stem. Set on E¹ E the half siding of the stem at E, which gives the terminating point. Next girth from the point E, in the Sheer, the positions of D, C, B, and A. Lay these distances off from number 4 frame along the centre line of the Half Breadth, and draw up perpendiculars 3^1, 2^1, 1^1, and produce level lines from B^1, C^1, D^1, E^1, the cutting points of the original frames with the harpin in its true position, until they meet the new perpendiculars, which are the true positions of the frames in the expanded form. A curve passed through these intersections is the expanded moulded edge. A yellow pine mould is made to this line, and the expanded position, number of the frames and ship marked upon it.

The Bevelled Edge.—Square down into the Half Breadth the point e, and on $e^1 e$ set off from the centre the half siding of the stem at e, which gives the terminating point. Lay a batten along the bevelled edge in the Sheer and mark on it the positions of a, b, c, d, e. Place the batten along the centre of the Half Breadth, keeping point a fair with 4 frame, and mark corrected position of the frames. Draw up perpendiculars from these points to cut level lines produced from the intersection of the frames with the bevelled edge in its true position. A curve through the spots will give the expanded bevelled edge, and the difference between this line and the moulded edge at any point, taken square to the latter is the amount of bevel placed on the bar.

The Bevelling Board.—Fig. 64. A piece of pine the width of the vertical flange of the harpin, having on the bevels taken at each frame is supplied with the mould. Holes are required to be punched in the vertical flange for temporary attachment to the frames. The position of the harpin is also marked on the frames and stem when they are turned; so that there is no difficulty in placing the harpin in its right position on the ship.

Fig. 64.

BEVELS HARPIN
4
3
2
1
STEM

Expansion of the Moulded Edge in the Sheer.—Girth the position of D, C, B, and A from E on the Sheer line E A, and set them along level line E W from E. Produce perpendiculars from the points, which give expanded position of the frames 1, 2, 3, and 4, on which set the level distances of the frames and half siding of the stem lifted from the Body. A line run through the spots gives expanded moulded edge.

The Stern Harpin is also made of angle bar, and appears a level line in the Body and Sheer plans. It is placed in between the rail and knuckle for securing the stern cants in position, until the after-deck stringer or bulwark plates are in place. Its extent is from two frames on one side to two on the other.

The Moulded Edge.—In **Fig. 65** the line A B and $a\,b$ is the moulded edge in the Sheer and Body. Square down into the Half Breadth from the Sheer the intersection of the moulded edge with the buttocks and centre line, and lift from the Body the distances c, d, b from a, and set them off in the Half Breadth on their respective frames. Draw line in, which will be moulded edge of the harpin in its true form, which is also the expanded form.

The Bevelled Edge.—Set off in the Sheer from A on the run of the stern and the buttocks and the frames in the Body the depth of the harpin bar, and square the points down from the Sheer into the Half Breadth. Lift the distances of the frames T, 1 and 2, from the centre of the Body on the bevelled edge and set them off in the Half Breadth on their respective frames. A curve drawn through the points will give the form of the bevelled edge. Show in the moulded edge of the cants in the Half Breadth; then the difference between the moulded and bevelled edges square to the first is the bevel on the vertical flange. A yellow pine mould is made to the moulded edge showing the positions of the centre line, cants, and frames T, 1 and 2. Bevels are lifted at centre, cants, and square frames, and placed on a piece of board the width of the vertical flange. Holes are punched by the frame-turners in the vertical flange of the bar for attaching the harpin to the cants and frames. The mould may be reversed on the boards for getting the shape of the opposite side of the ship. This harpin is not used in all yards.

NOTE.—These harpin bars do not of necessity require bevelling, except in extreme cases. Of course, where the cants do not extend so high, it should be placed lower down.

Fig. 65.
STERN HARPIN.

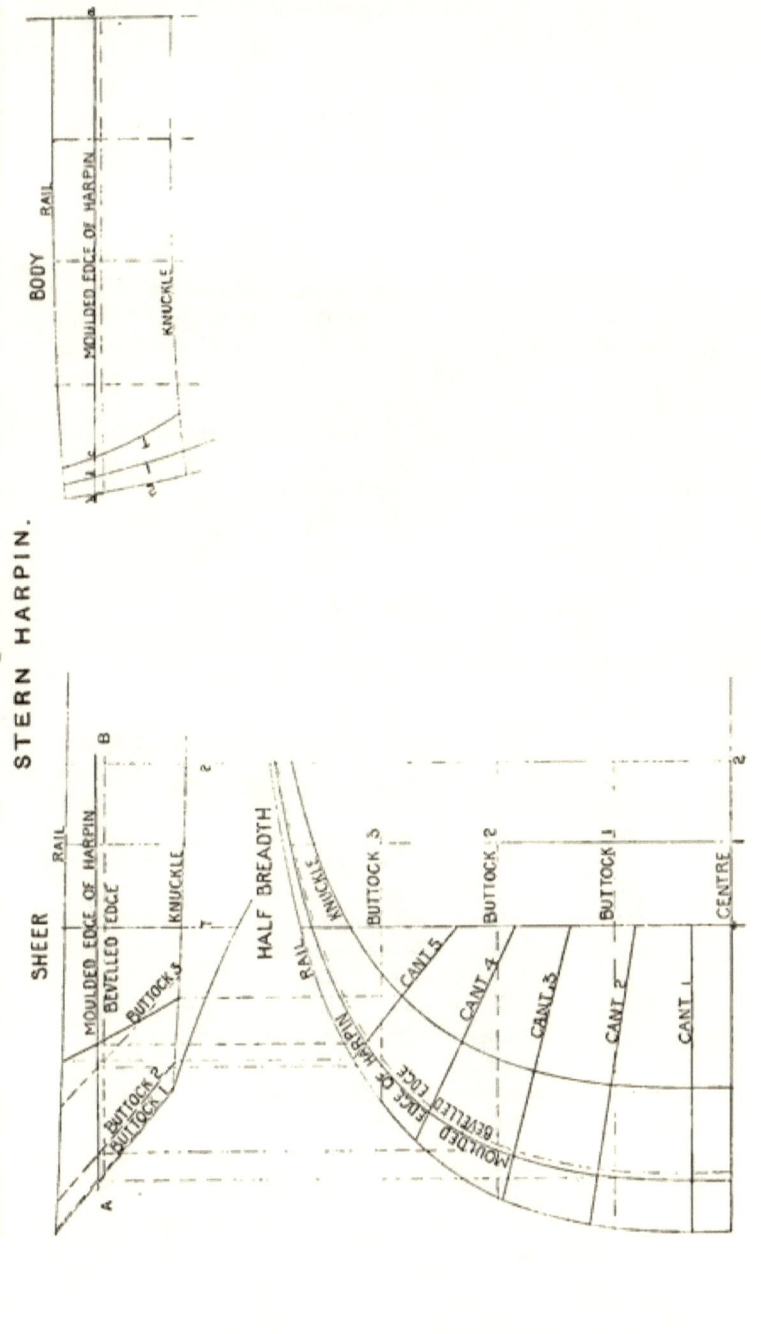

CHAPTER VIII.

Principal Moulds, and the order they are sent into the Yard—Stem—Stern Frame—Shaft Struts—Stern Tubes—Flat Plate Keel—Centre through Plate Keelson—Boat Beams.

MOULDS.

Wood moulds, or templates, are made in the loft of various parts of the vessel for the use of the platers and others, for forming the material into its proper shape. The principal of these, and the order in which they should be sent out, is as follows:—

1. Flat keel plate.
2. Centre keelson plate, with templates for vertical bars.
3. Beam camber, with templates for knees.
4. Stem and stern post club-foot sections.
5. Struts and stern tubes in twin screws.
6. Stern cants.
7. Stem.
8. Stern, common, and sheer harpins.
9. Stern expansion and set moulds.
10. Stringer plates.
11. Cargo hatches.
12. Deck-houses.
13. Clipper stem lacing piece, bowsprit, trail board, and figure block.
14. Boat beams.
15. Hawse-pipes.

Nos. 1, 2, 4, 5, 7, and 14, are described in this chapter; and Nos. 3, 6, 8, 9, 10 to 13 and 15, on pages 31, 32, 69, 50, 88 and 96, respectively.

The Stem.—In mercantile vessels it is made of solid iron, of rectangular section. Where a flat plate keel is associated, the foot usually has a club, as shown in **Fig. 66**, with a recess on the lower edge for the scarph of the keel. When the ship is faired-up on the loft floor, section moulds are given to the forge to form the foot. A flat mould of pine (shown by Fig. 67) is made, in the meantime, to the inside edge of the stem for turning the bar to its shape on the yard bending slabs. This mould has marked on it the position of the keel scarph (approximately), frames, shell plate landing edges, ribbands, decks, and a level or water-line on, or about, the load draft. Before turning the bar, the position of the plate or bar keel ending on the building blocks, should be lifted relative to an adjacent frame, and transferred on to the mould. It will be readily seen that, to secure the *designed* length, this should be carefully done: besides, any mistake would affect the end shell-plates. Where the stem, above the forefoot, is not square to the keel base, a perpendicular is lined on the mould and transferred to the stem after turning, beside the position

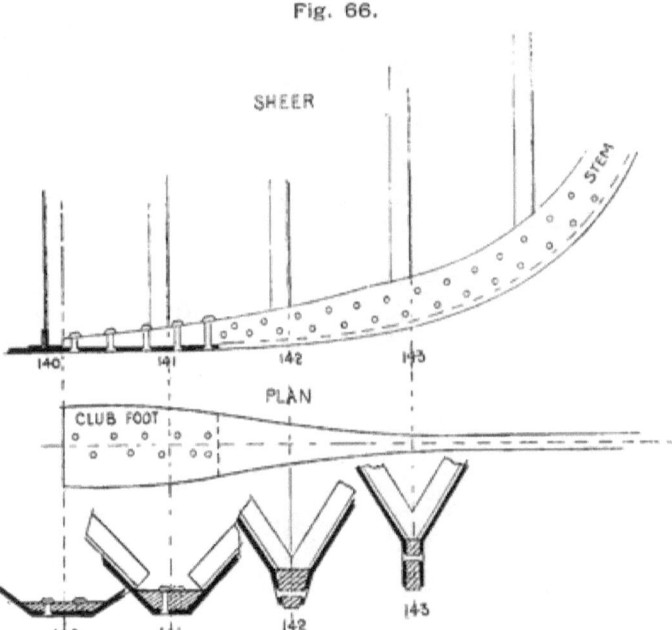

Fig. 66.

of the shell-plate edges, decks, and ribbands. The height of the level line is centre-punched on the fore edge of the bar; which will be found useful in erecting. In some yards the rivet holes are drilled in the bar previous to turning, leaving out those in way of the landing edges until the ship is faired-up on the blocks. The better plan is considered, where convenient, to drill these holes after the bar is in form.

In some cases, instead of the club-foot, an angle bar, or flanged plate, is tapped on to the stem bar to take the keel and garboard strakes. This is shown in Fig. 68.

Fig. 67.

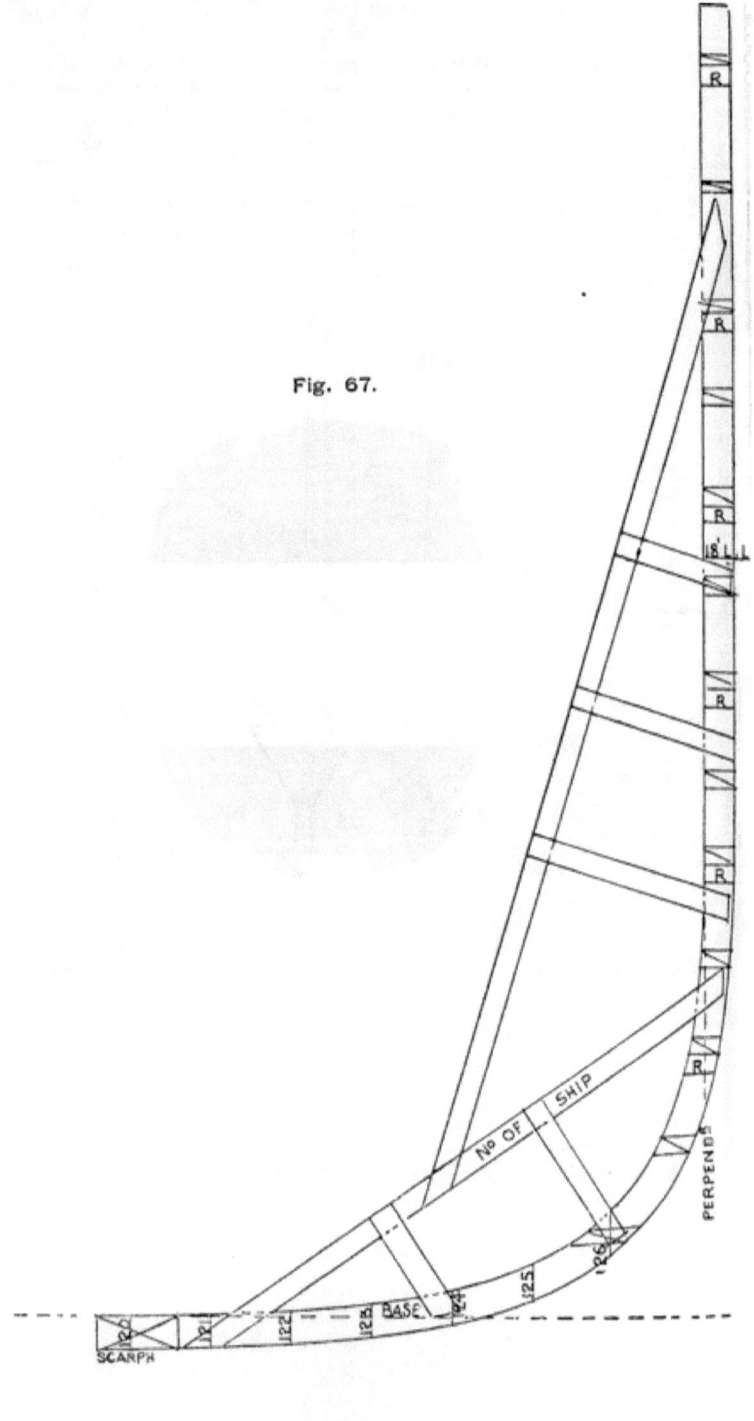

The form of the stem scarph, when associated with "side bar keel," is given in **Fig. 69**.

Stern Frame.—The forge usually makes its own moulds from a complete figured drawing on a scale of half an inch to a foot, which is supplied by the shipbuilder, together with enlarged detail figured sketches of riveting, gudgeons, boss, etc. In the case of a flat keel a

Fig. 68.

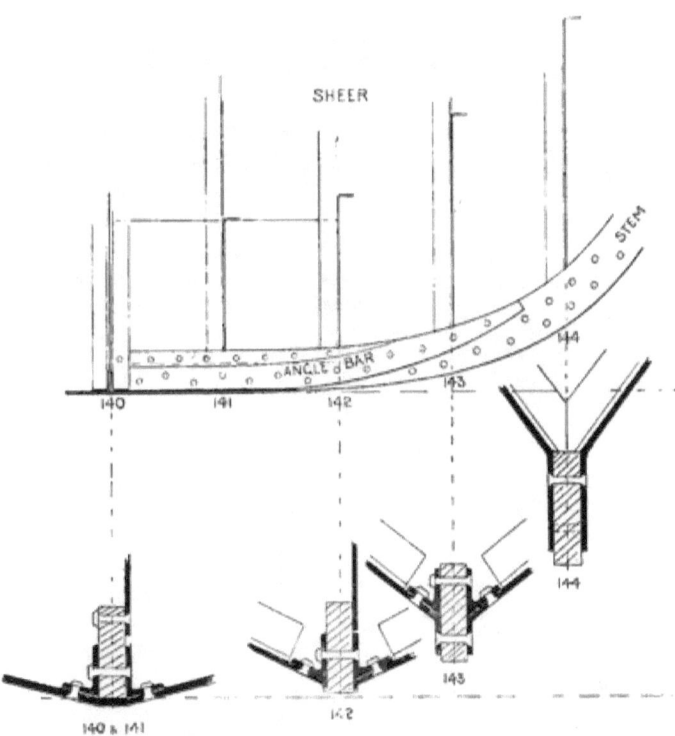

club-foot is fitted in most yards, and for the purpose of forming this, after the ship is faired-up on the loft floor, section moulds are supplied to the forge.

Shaft Struts or After Brackets in Twin Screws.—They are usually made of cast steel. For the purpose of accuracy it is the custom for the loftsman to make a template or drawing, **Fig. 70**, from the loft floor lines, through the athwartship centre, showing the

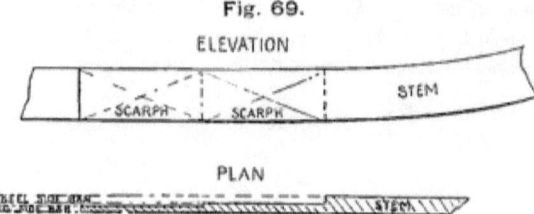

Fig. 69.

spread of the legs, boss, and rough bore. This is tried upon the ship when in frame, before it is sent to the works, from which, in conjunction with a figured drawing or tracing, moulds for casting are made. It may be necessary to give bevels for shafts. Where the palms are planted on the outside of the shell plating, sectional templates of the ship's side should be supplied to form patterns. A template of T form is generally made, showing the character of the shell.

Stern Tubes in Twin Screws.—There are many ways of finishing off the apertures for the shafts in the sides of a vessel. A simple and inexpensive method is shown on page 13. In that case a cast steel hollow ring, reproduced on a larger scale in **Fig. 71**, is

MOULDS. 79

attached to the framing D. The shell plating B and B¹ is lapped round the fore and aft rim A, and connected with a substantial number of rivets. On the after side of the ring there is a projecting flange C and C¹, with a rabbet to take shell plating. To cast these rings, a full sized section is laid down on a board (see E F), and a view of the after and forward sides of the ring drawn alongside. This is sent to the works, who make their own moulds. In vessels of larger power this ring may be lengthened (see **Fig. 72**), and additional projecting pieces a and b made for taking partial bulkheads. The shell plating laps on to the ring at d and d^1, and the sides have a web c and c^1 to take the plating. For making moulds, for casting, a full sized drawing on a board is sent to the works showing longitudinal

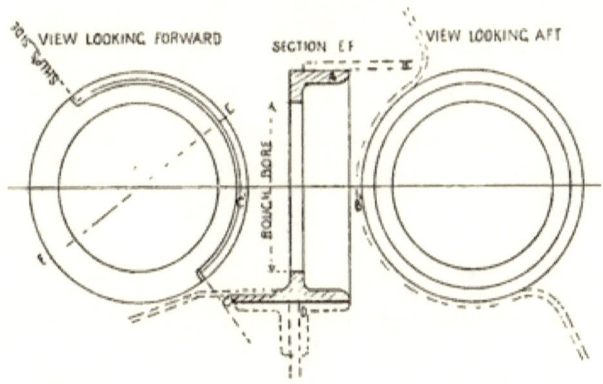

Fig. 71.

STERN TUBE RING.

and cross sections. Skeleton moulds would be better tried on the ship before being sent away.

Sometimes these rings are forged.

Flat Plate Keel.—Two small scrieve boards, about 3 feet square, are supplied to the yard, showing the form of the bottom at each frame for the extent of the keel. Those which are shown in **Figs. 73** and **74** are given in conjunction with the wood template, **Fig. 75**, having longitudinal pieces A and A¹ the width of the seams, the outside edges representing the expanded breadth of the plate, with pieces nailed across showing the shell flange of the frames. The heavy line in the sketch being the heel of the bar, which is planed to distinguish it. End pieces D, nailed on, show the position and width of the laps, or half the butt strap when flush. Care should be taken to get the sides of the template perfectly parallel, and the ends square, to the centre. Then the centre lines for the edge holes and centre bars may be struck in the full length, and on the crosspieces for the frames. A piece of board is now got perfectly rectangular, the width of the plate and the length of the butt lap, upon which is divided off

Fig. 72.

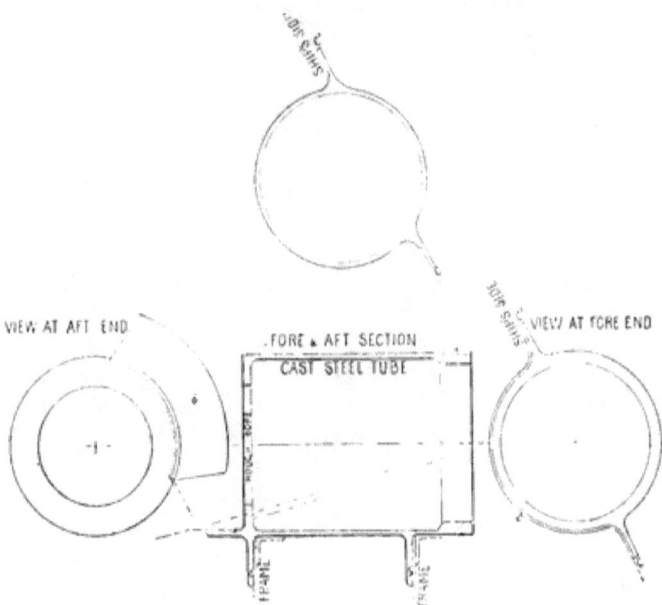

the rivet holes, as shown in **Fig. 75**, to take seams and centre bars. This is placed fair upon one end of the template and tacked in position, and the holes bored through it and the template at the same time. It is then taken off and holes marked off from it on the other end of the template, after which the longitudinal seam and the frame holes are divided in and bored out. It will be noticed that only one hole is placed through each seam in way of the frames, which is next to the outside sight edges of the keel plate for caulking purposes. All the keel plates may be marked off from this template, while the set on the plates is lifted from the scrieve boards, **Figs. 73** and **74**.

A batten template of the holes should be given to the frame turner in order that those in the frames may be made to agree.

Fig. 73.
AFTER BODY.

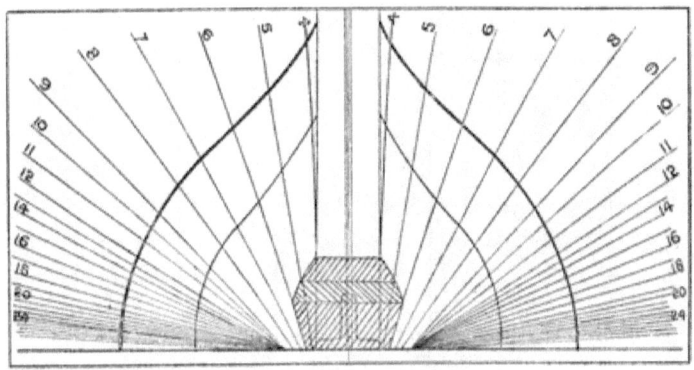

Fig. 74.
FORE BODY.

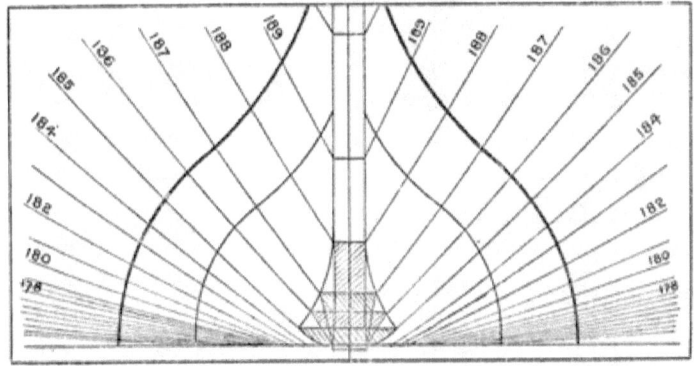

In some cases it is best to lift the set of the foremost and aftermost plates from the ship, especially that part in way of the club-foot of the stem and stern posts.

Fig. 75.
KEEL

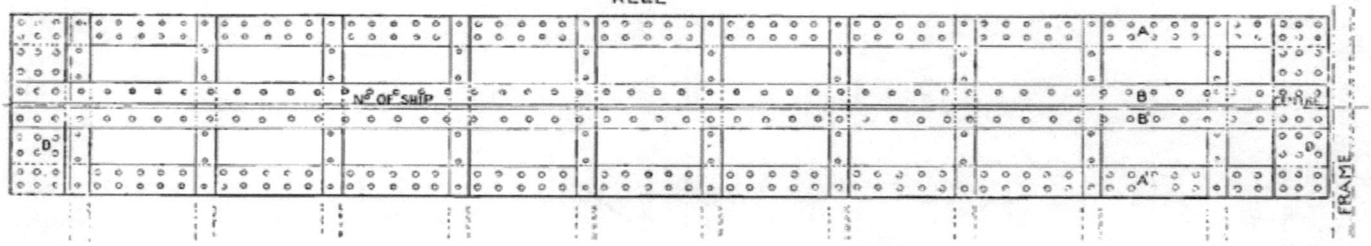

Fig. 76.
KEELSON

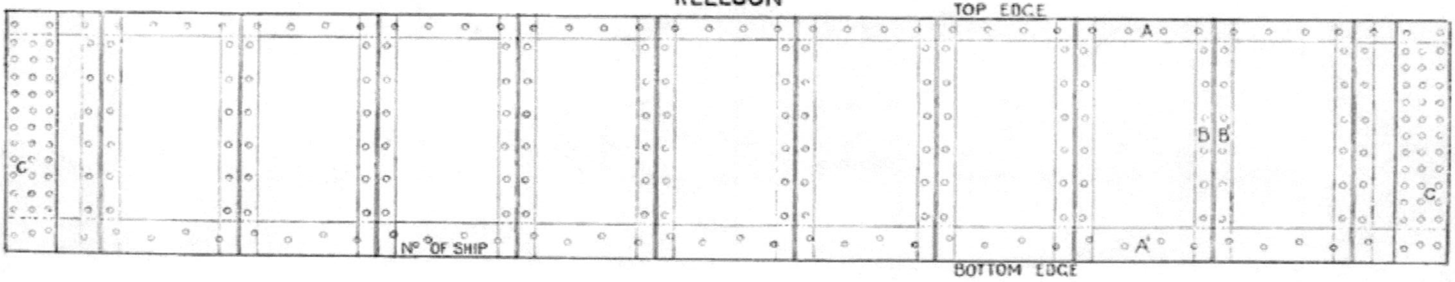

Centre through Plate Keelson.—The plates are more easily worked when of uniform length. Two duplicate wood templates are made for marking off in the case of a flat plate keel, and one with a side bar keel.

Fig. 76 shows such a template, the lines for which are marked off on the loft floor. The top edge is parallel to the bottom, the longitudinal pieces A and A¹ being the width of the vertical flanges of the top and bottom bars respectively, to which pieces B, B¹ are nailed, showing the position and width of the vertical bars, with the edges, representing the heel, planed. Pieces C show the position and width of the butts. The holes in the longitudinal bottom bars are slightly recled. A piece of board, perfectly rectangular, is made the width of the plate and the length of the butt, upon which the rivet holes are arranged to take the top and bottom bars: this is tacked into position on the end of the template, and the holes bored through both. It is then lifted and transferred to the other end, and the holes marked off and bored. The top and bottom and vertical bar holes may now be divided into the regulation spacing and bored.

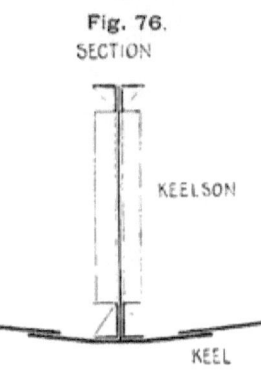

Fig. 76.
SECTION

KEELSON

KEEL

A duplicate template may be marked off by laying one on top of the other, or tacking temporarily two templates together in the first case, and boring holes through at the same time.

A separate strip template is given to the apprentice platers, showing the length of and holes in the vertical bars B, B¹, and another template showing those in the other flange attached to the floor. A copy of the latter is also given to the board-men, to punch floors to receive vertical bars.

In the case of a centre through plate keelson with a side bar keel, the template is only made to the top of the side slabs, the plate being punched above the keel first and afterwards attached to side slabs, and the three thicknesses drilled through. The holes in the keel are set off by the loftsman to the regulation distance. It is best to make a short template the width of the keel, with the holes bored through, that can be shifted as each plate is drilled.

Boat Beams.—Draw in on the loft floor the form of the ship's side in way of each beam. In **Fig. 77** fix the height A at the centre line, such that you have comfortable head room at the sides below the web of the beam. Draw in camber line A B and connect it at **E** with side, C D produced. A skeleton wood mould is made to the line C D B A, showing the correct length and the centre. Cross battens to keep it in form are fitted from D to A, and from D A to E B A edge. It may be found necessary where there is considerable difference of the vessel's side to make several moulds. Sometimes these beams

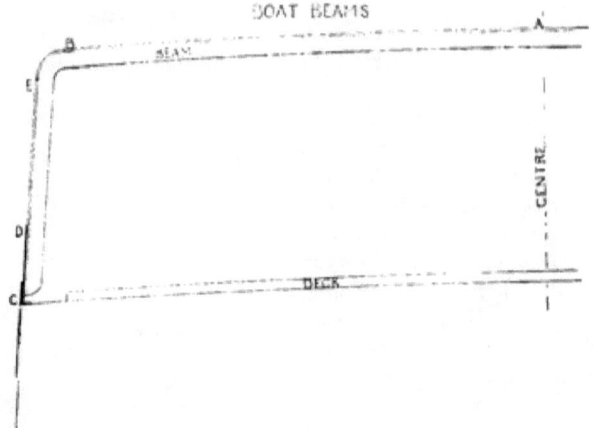

Fig. 77.

only extend to the deck erections, in other cases they are butted at the centre, or a little to the sides. In any case a half mould ought to answer the purpose, if the centre is correctly marked for each beam.

CHAPTER IX.

Poop Round: Obtaining Lines and Fairing-up—Expansion of Plating. *Turtle Back:* How to Obtain and Fair-up the Form, Expansion—Plate Edges.

POOP ROUND AND TURTLE BACK.

The majority of sailing ships and steamers are fitted, above the upper deck, with after and forward short erections for the berthing of the crew, and for protection from seas. That aft is called a poop, and forward a forecastle. To prevent these erections from giving a heavy appearance to the vessel, they are usually rounded away at the sides above the main rail. In many cases, especially in West of England built vessels, the forecastle assumes a turtle back form. The method of deciding and working out the form, and other points, of these will now be described.

To Obtain the Lines and Fair-up a Poop Round.—In Fig. 78, draw in the Sheer the poop deck at the centre line A B; and through the point C, the top of the rail, describe the quadrant of a circle to meet the line A B at B. This line C B A is the form of the poop at the centre line. Set up level lines $a\,b$ and $c\,d$, about 9 inches apart, above the rail height C to cut the round. Square down on to the Half Breadth centre line the intersection of these level lines with the circular part C B, and the position of B or poop angle bar. Run lines $e f$, $g h$, and B¹ E, from these points, parallel to the rail. Then square up on to the rail in the Sheer the position of the buttocks, and set 9 and 18 inch sheer lines above these points. Project up from the Half Breadth on to their respective sheer lines the intersection of the buttocks and points on B¹ E. Place the sheer lines in the Body above the T, 1, 2, and 3 section rail heights. Lift from the Half Breadth the positions of $e f$, $g h$, and B¹ E on the T, 1, 2, and 3, and lay them off on corresponding sheer lines at each body section square out from the centre. Next mark off, on the centre line of the body, the deck centre heights on the T, 1, 2, and 3, and draw in curve of the beam in each case from camber curve given in the figure ; and connect the curve with the corresponding spots on the sheer lines, set off for the round, maintaining the height of B above C C¹ at the angle bar. This will give you the form of the poop on T, 1, 2, and 3, which may be checked by transferring the heights above the level line on each buttock into the Sheer, and connecting them by fair curves with the spots on the round, already projected from the Half Breadth. If these lines are faired in each plan, as you proceed, it will be found that sections taken square to the rail in the Half Breadth are circular, corresponding with the centre line C B. Cant No. 4 is shown projected into the Sheer: the height on T may be found by setting off T D in the Body on the transom camber curve, and lifting the point, D^1 E, from the level line into the Sheer T D^2. The formation of a poop side of a flatter form is found in the same way, after the centre C B is settled.

Expansion of a Poop Round.—In Fig. 79 show the cants in position in the Half Breadth, and arrange the butts of the plating to clear. Girth the rail in the Half Breadth for the position of the moulded edge of the cants, square frames T and 1. Set these off along the expansion level line, and erect perpendiculars from the points, on which lay off respective sheer heights, lifted from the Sheer above the level line—that is, the centre, cants, frames T and 1, and draw a curve through the spots. This is the bottom edge of the rail line expanded: the top edge of the rail may be drawn parallel to it, the depth of the moulding. Girth the centre line section in the Sheer, from the top of the rail, for the position of a, b, c, and d, and put these distances out on the centre line in the expansion A^1, a^1, b^1, c^1, and d^1, and line in curves a^1, b^1, c^1, etc., through the spots parallel to the rail. In the plan show in lines ef, gh, kl, mn, op, square to the rail and about the centre of each plate. Then girth the rail for their position, and transfer them into the expansion along the rail A^1 A^2, and from the points erect perpendiculars to the rail curve; the correction for the slight sheer of rail can be made if considered desirable. Girth in the Half Breadth the width of each plate to its butt on each side of these centre lines, along the rail, sheer lines, angle bar, and inner edge, and set them off in the expansion on each side of the centre lines $e^1 f^1$, $g^1 h^1$, etc., tracing curves through the spots as shown by heavy lines, which will give the true form of the plate in each case. In ordering, allow for "sny," also round and hollow at the ends. Care must be taken, when fixing the point d, that the plate is sufficiently wide to take the margin plank and deck ends.

Turtle Back.—In Fig. 80, draw down the form of the rail in the Sheer, Half Breadth, and Body, for the extent of the turtle back; and line in the position and form of the buttocks in each plan from level line to the rail, also the frames T, 1, 2, and 3; putting in A, B, C, and D temporary sections abaft of the transom; then you are ready.

5 —

5 —

75 — LEVEL LINE

CA —

GAIN —

GAIN —

1 —

1 —

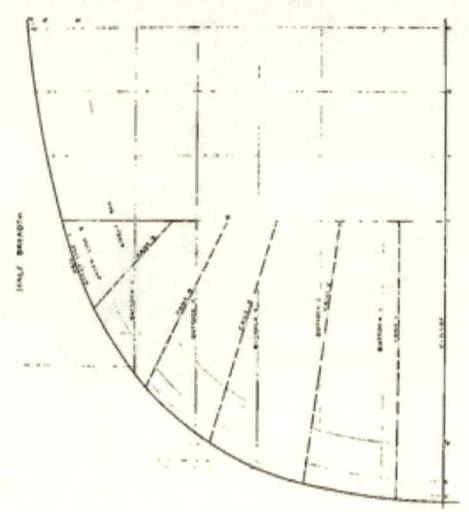

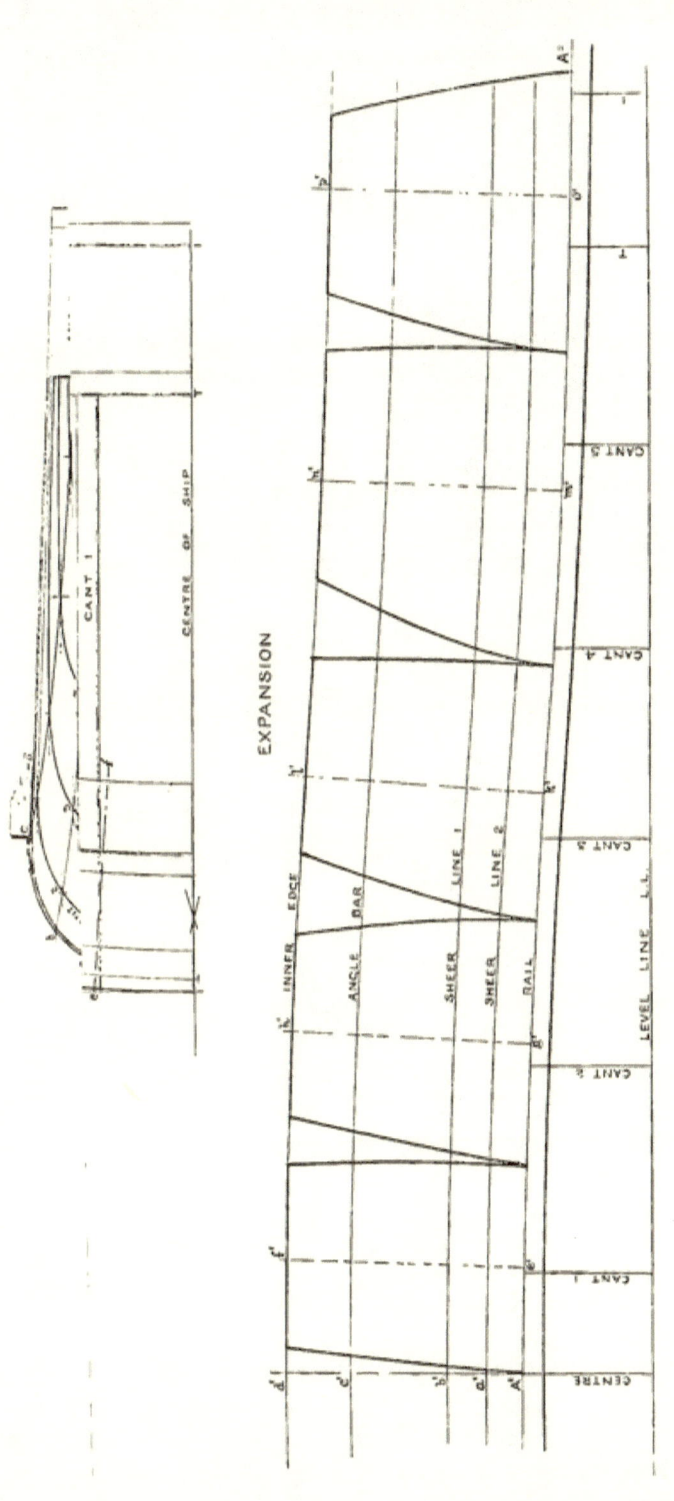

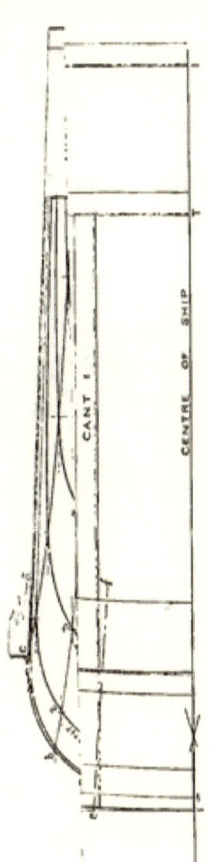

EXPANSION

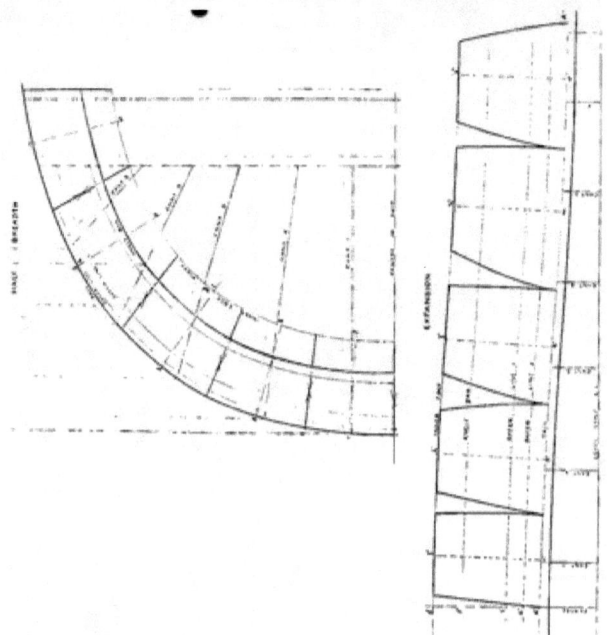

Fig. 80.

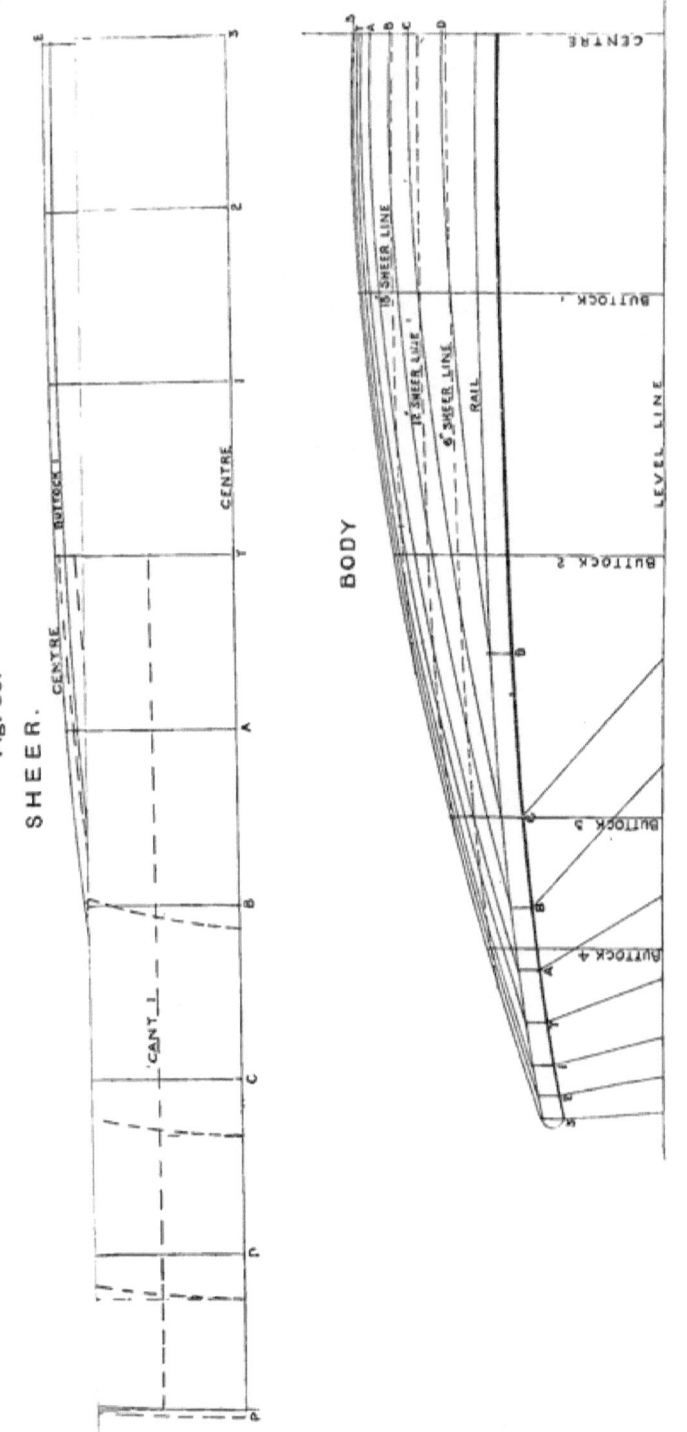

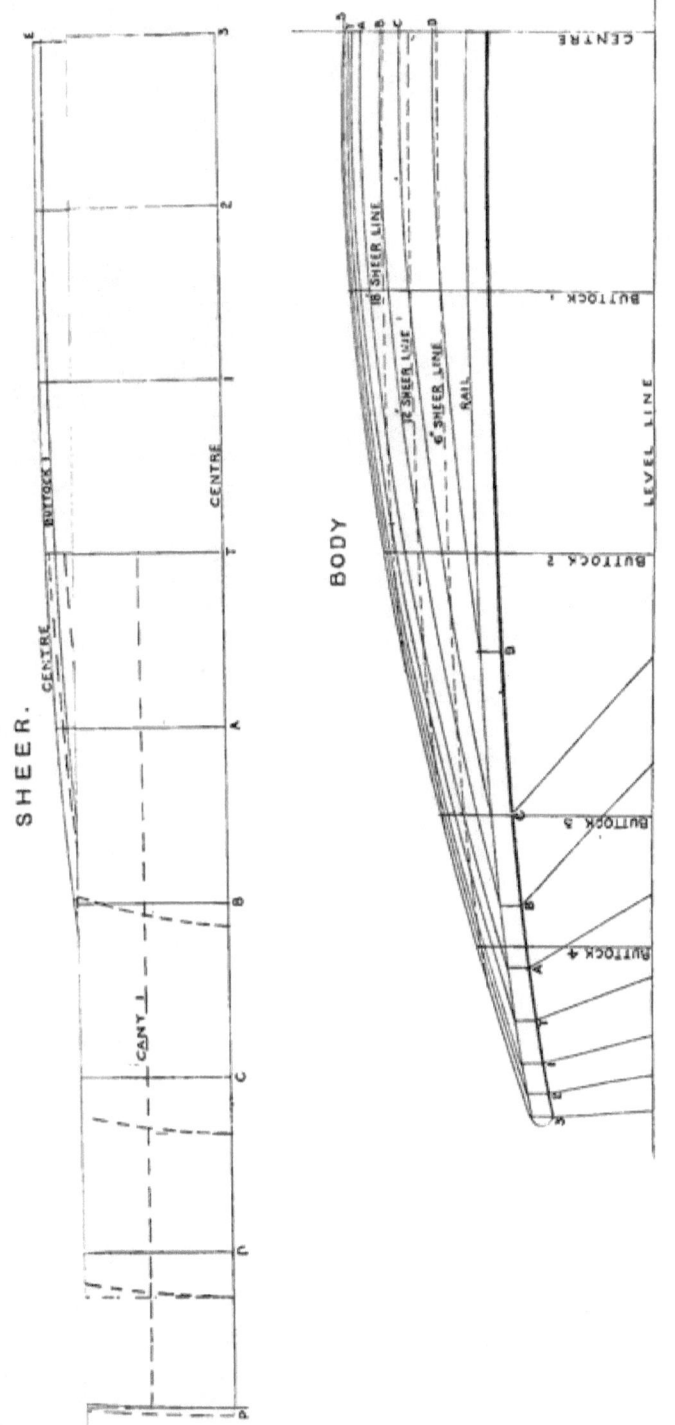

Fig. 80.

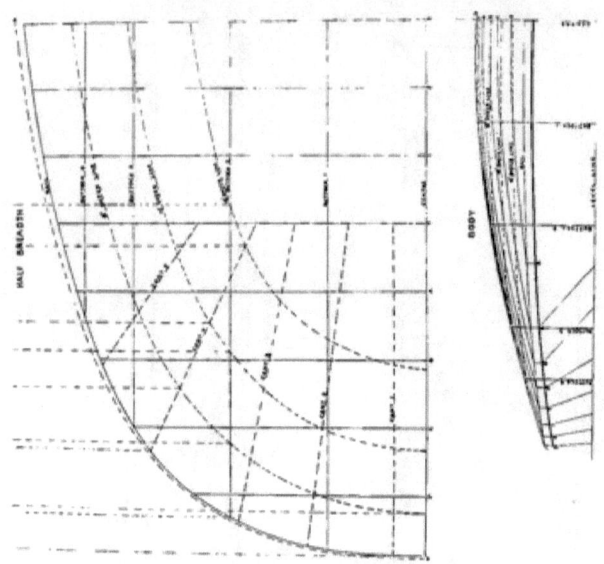

To Obtain Lines for the Turtle Back.—Settle the height, E, from the upper deck, and the form of the line, E A, at the centre of the ship; lift the heights on each section above the level line, and transfer them on to the centre in the Body above the corresponding level line. Draw in the Body the required form of the cross section on 3 frame, usually a fuller line than the ordinary camber curve explained on page 33, also the transom cross section, about the same curve as No. 3 frame. Transfer into the Sheer, from the Body above the level line, the cutting points of 3 and T on alternate buttocks, and run fair line to the terminations on the rail; somewhat like the centre, only gradually spread towards the after end. Make any correction you find necessary for fairness in the Body and Sheer before running the other buttocks; then the intermediate cross sections can be readily drawn in by lifting the sheer heights on the buttocks and transferring them into the Body.

Sheered lines may be used for obtaining and fairing form in conjunction with the buttocks. They are first placed, as shown in the Sheer, about 6 inches and 12 inches parallel to the rail; their intersection with the buttocks are squared down into the Half Breadth. More points may be got by placing the sheered lines in the Body, and transferring widths on the frame stations into the Half Breadth. These lines would enable you to place an additional check upon the previous fairness.

The cants are shown in the Sheer dotted, projected from the Half Breadth in conjunction with the heights on the transom lifted from the Body.

To Expand the Form of a Turtle Back.—It is considered sufficiently accurate to girth the centre line E A in the Sheer for the position of the frames, and the point A from E; and set them off from 3 on the centre line of the Half Breadth, from which points erect perpendiculars, and on them lay off the girthed widths of the cross sections. It is necessary to girth the front section No. 3 for the position of the buttocks, and place them in the Half Breadth parallel lines to the centre, on which are plotted off the girthed lengths of the sheer buttocks from 3 frame. This will give you sufficient spots to draw in the curve of the expanded form of half of the turtle back P K. Allowance should be made at the outside edges for connection to other plating, usually 3 inches round the sides and the width of the beam flange at the fore end. The plate edges are made parallel to the centre line, and the cants and butts arranged, after which the plates may be measured off and ordered.

To Obtain Lines for the Turtle Back.—Settle the height, E, from the upper deck, and the form of the line, E A, at the centre of the ship; lift the heights on each section above the level line, and transfer them on to the centre in the Body above the corresponding level line. Draw in the Body the required form of the cross section on 3 frame, usually a fuller line than the ordinary camber curve explained on page 33, also the transom cross section, about the same curve as No. 3 frame. Transfer into the Sheer, from the Body above the level line, the cutting points of 3 and T on alternate buttocks, and run fair line to the terminations on the rail: somewhat like the centre, only gradually spread towards the after end. Make any correction you find necessary for fairness in the Body and Sheer before running the other buttocks; then the intermediate cross sections can be readily drawn in by lifting the sheer heights on the buttocks and transferring them into the Body.

Sheered lines may be used for obtaining and fairing form in conjunction with the buttocks. They are first placed, as shown in the Sheer, about 6 inches and 12 inches parallel to the rail; their intersection with the buttocks are squared down into the Half Breadth. More points may be got by placing the sheered lines in the Body, and transferring widths on the frame stations into the Half Breadth. These lines would enable you to place an additional check upon the previous fairness.

The cants are shown in the Sheer dotted, projected from the Half Breadth in conjunction with the heights on the transom lifted from the Body.

To Expand the Form of a Turtle Back.—It is considered sufficiently accurate to girth the centre line E A in the Sheer for the position of the frames, and the point A from E: and set them off from 3 on the centre line of the Half Breadth, from which points erect perpendiculars, and on them lay off the girthed widths of the cross sections. It is necessary to girth the front section No. 3 for the position of the buttocks, and place them in the Half Breadth parallel lines to the centre, on which are plotted off the girthed lengths of the sheer buttocks from 3 frame. This will give you sufficient spots to draw in the curve of the expanded form of half of the turtle back P K. Allowance should be made at the outside edges for connection to other plating, usually 3 inches round the sides and the width of the beam flange at the fore end. The plate edges are made parallel to the centre line, and the cants and butts arranged, after which the plates may be measured off and ordered.

CHAPTER IX.

Expansion of a Stringer Plate with no Sheer and with Sheer—Template—Allowance for Knees in Ordering Tee Beams, and in Bulb Plates.

EXPANSION OF A STRINGER PLATE AND BEAM KNEES.

Expansion of a Stringer Plate.—Sometimes the stringer plate edges are expanded on the loft floor for the purpose of making wood templates for the platers. It is sufficient, where the deck at the side is practically level, to set off the spacing of the beams, and on them from the side curve to lay off the widths of the plate : but when the deck has considerable sheer, like **Fig. 81**, it is necessary to girth the sheer line for the position of the butts and beam points a, b, c, d, e, f, g, h, and k, and lay them out on a level line A B, giving points a, b^1, c^1, d^1, etc., which extend square to the centre line over the stringer. Produce level lines through the intersection of the outer edge of the stringer plate with the original beam lines, until they cut the expanded beam lines at b^2, c^2, d^2, etc., which intersection gives points for the expanded edge of the stringer, shown in the figure by a dotted line. From this edge on the expanded beams set off the width of the stringer plate, and draw inner edge through the spots : and then fix the butts. The figure C D E F is the expansion, for which a wood skeleton mould is made showing edges, butts, and beams in the usual way. The holes for the beams are generally lifted from the ship by applying template in position, and marking from the underside on the crosspieces with a circular piece of wood dipped in liquid whitening. Those for the edges and butts are divided in along the side battens. The holes for beams may be marked from the beam battens, which were used for lining off the holes in the beams.

To Find the Allowance for Beam Knees or Arms.—The Classification Registry rules require that the *depth* of beam knees should be twice and a half the depth of the beam web, and once and a half the depth across the throat. For special reasons it is sometimes more. Knees to tee beams are best formed by sawing up the depth, and turning the lower part down to form the knee : a piece of plate being welded into the angular vacancy. In ordering, when done in this fashion, allowance for turning down has to be made over the

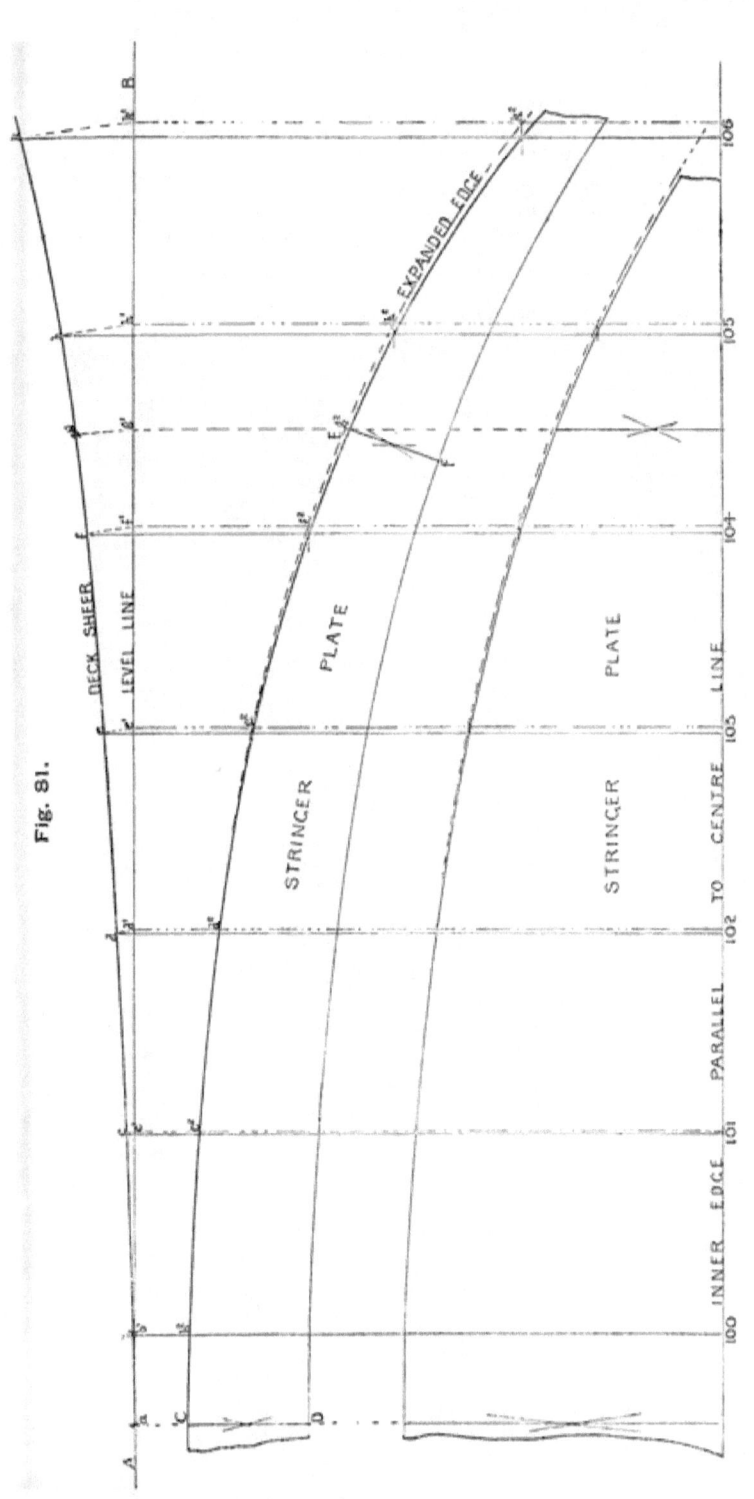

Fig. 81.

width of the deck plan. In **Fig. 82** the tee beam knee is shown in position. Girth the centre line ab of the turned down part, and set it out straight $a^1 b$, then cd is the extra allowance for one side of the ship, which works out in practice about the depth of the beam over the half length.

By **Fig. 83** it will be seen that, at the ends of the ship, less than the midship allowance would do owing to the fall in of the side. The

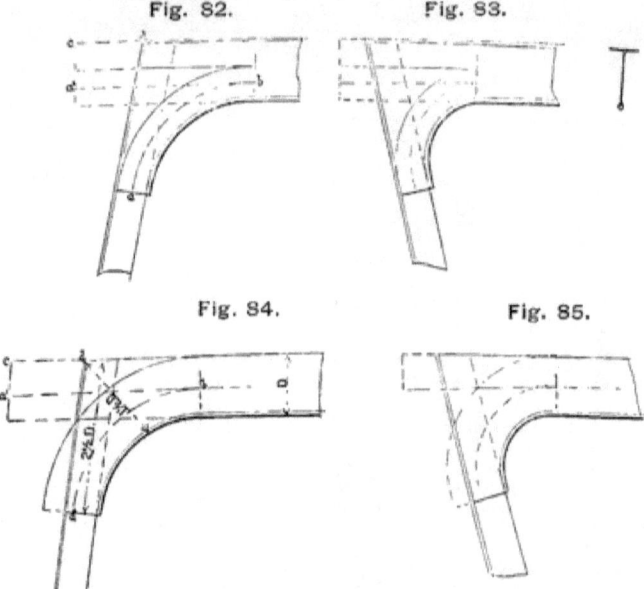

Fig. 82.   Fig. 83.

Fig. 84.   Fig. 85.

wisest course is to draw a line in conjunction with the deck plan, showing the allowance at each beam. This line may be easily found by trying a few sections.

Fig. 84 shows a bulb plate beam which is turned down bodily, and a piece welded in the top corner. Draw in the centre line ab. Girth this line, and set the distance, $b a^1$, out straight. Then cd is the allowance for each side. The end allowance may be reduced (see **Fig. 85**).

NOTE. — It may be mentioned that the difference between the level width of a deck and the girthed width, allowing for the ordinary camber, is about one inch on the midship beam.

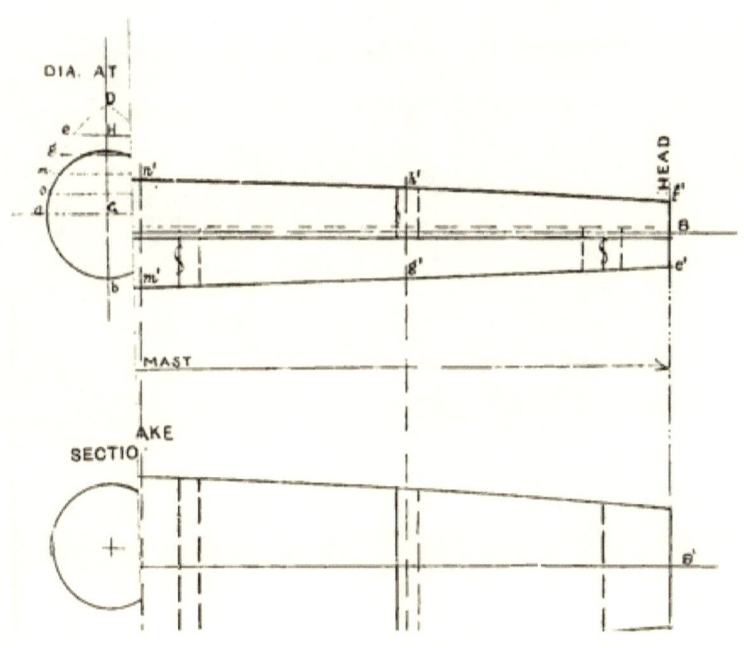

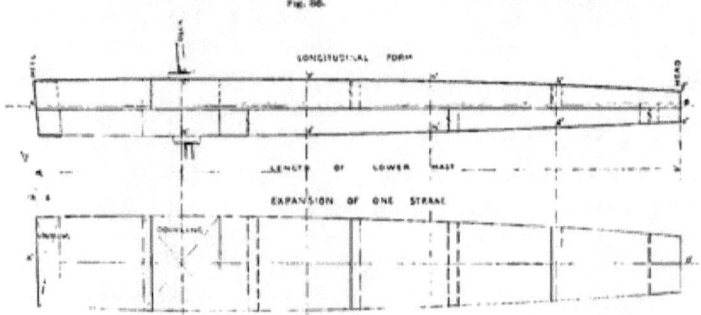

Fig. 66.

CHAPTER XI.

How to Obtain the Form—Expansion—Doubling at Deck—Doubling at Heel— Mast Tube Expansion.

IRON AND STEEL MASTS.

Masts are, in the most of cases, tapered and raked aft, the rake varying from $\frac{1}{2}$ to $2\frac{1}{2}$ inches.

To Obtain the Form.—The diameter at the head is about five-eighths and the heel about three-fourths of the deck diameter. Draw down in **Fig. 86** a straight line the length of the lower mast, on a scale of $\frac{1}{2}$ inch to the foot. Let this line A B be the centre, and on it set off the correct positions of the heel, deck, and head. The housing is best measured from the Sheer Draught, making allowance for camber of beam. On A B, produced clear of the heel, describe the circle $a\,b\,c$, with diameter equal to that required at the deck. From the points a and c describe arcs a D and c D with radii $a\,c$. Find a point $e f$, between the arcs, equal to the decided diameter at the head. Divide G H into a suitable number of equal spaces, say four, and level lines through the points to the arcs. Divide the distance on the line A B, between deck and head, into the same number of equal spaces as G H, and square points across the plan. Lift from the section, $a\,c$, $o\,p$, $m\,n$, $g\,h$, and $e f$, around G H, and set them off on their corresponding lines, $a^1 c^1$, $o^1 p^1$, $m^1 n^1$, $g^1 h^1$, and $e^1 f^1$, divided equally on each side of the centre: treat the part below the deck in the same way. Pin a stiff flat batten to these spots, and run lines in. If the mast is raked, set down a foot on perpendicular A X, and place the rake per foot on perpendicular X Z; and from this point Z draw Z A, which gives the cutting line for the heel. The head is made square to the centre line. This is the usual method, but some shipbuilders prefer a curve with less pronounced round.

Expansion of the Mast.—It is only necessary to expand one strake, because all the strakes are the same form. Take a mast with two plates in the round, which is the most common. It is evident that each strake, if overlapped at the edges, will require to be wider at any point than half the circumference by the width of one lap. Calculate the circumference due to the diameter at each of the points

CHAPTER XI.

How to Obtain the Form—Expansion—Doubling at Deck—Doubling at Heel—Mast Tube Expansion.

IRON AND STEEL MASTS.

Masts are, in the most of cases, tapered and raked aft, the rake varying from $\frac{1}{2}$ to $2\frac{1}{2}$ inches.

To Obtain the Form.—The diameter at the head is about five-eighths and the heel about three-fourths of the deck diameter. Draw down in **Fig. 86** a straight line the length of the lower mast, on a scale of $\frac{1}{2}$ inch to the foot. Let this line A B be the centre, and on it set off the correct positions of the heel, deck, and head. The housing is best measured from the Sheer Draught, making allowance for camber of beam. On A B, produced clear of the heel, describe the circle $a\,b\,c$, with diameter equal to that required at the deck. From the points a and c describe arcs a D and c D with radii $a\,c$. Find a point ef, between the arcs, equal to the desired diameter at the head. Divide G H into a suitable number of equal spaces, say four, and level lines through the points to the arcs. Divide the distance on the line A B, between deck and head, into the same number of equal spaces as G H, and square points across the plan. Lift from the section, $a\,c$, $o\,p$, $m\,n$, $g\,h$, and $e\,f$, around G H, and set them off on their corresponding lines, $a^1 c^1$, $o^1 p^1$, $m^1 n^1$, $g^1 h^1$, and $e^1 f^1$, divided equally on each side of the centre; treat the part below the deck in the same way. Pin a stiff flat batten to these spots, and run lines in. If the mast is raked, set down a foot on perpendicular A X, and place the rake per foot on perpendicular X Z; and from this point Z draw Z A, which gives the cutting line for the heel. The head is made square to the centre line. This is the usual method, but some shipbuilders prefer a curve with less pronounced round.

Expansion of the Mast.—It is only necessary to expand one strake, because all the strakes are the same form. Take a mast with two plates in the round, which is the most common. It is evident that each strake, if overlapped at the edges, will require to be wider at any point than half the circumference by the width of one lap. Calculate the circumference due to the diameter at each of the points or at the butts, and half the quantity, to which add the width of one lap. Set off half of these amounts on their respective points, at each side of the centre line $A^1 B^1$, drawn parallel to A B, and run curves through the spots. Make allowance for the length of the heel in each strake due to the rake. This will be the expansion of one strake of the plating, and seeing that the other side is the same, arrange the butts for both strakes on this, clear of each other and of the deck, with plates of equal length, except the ends; and of such a length that they can be got into the special yard rolls. Order the plates with allowance for any round on the edge of the expansion, etc.

Doubling in Way of the Deck.—The inside strake can be doubled on the outside, and the outside strake on the inside from edge to edge, and about equal lengths below and above the deck; the extent of the doubling is settled by the Classification Society.

Doubling on the Heel.—Seeing this is usually parallel to the foot it requires expanding to get its true form. **Fig. 87** shows the doubling on the foot of a mast with considerable rake and taper. Draw in the centre line $m n$ and $e a$, $s o$, and $f l$ square to it. Then describe a circle from A equal to $e a$, and from B equal to $f l$. Divide the circumference of each into 8 equal parts, and produce the points on A to $e a$, and on B to $f l$, parallel to $m n$. Join each set of points, top and bottom, and produce the lines on to $e t$ and $f u$. From the point S draw level line $q^2 q^2$ equal in circumference to $s o$, and divide it into 8 equal parts. Erect perpendicular $m^1 n^1$. Then set off $c^2 c^2$ and $h^2 h^2$, with vertical distances from $q^2 q^2$ equal to $s e$ and $s l$ respectively, and set on $c^2 c^2$ about $m^1 n^1$ the circumference of $e a$, and on $h^2 h^2$ the circumference of $l f$. Divide each into 8 equal parts, and join the corresponding spots on the three lines, which produce above $c^2 c^2$ and below $h^2 h^2$. Measure the distances of the top of the doubling plates, in **Fig. 87** above $a e$ on the run of the divisional lines, and set them on their respective lines in the expansion above $c^2 c^2$, and those for the bottom below $h^2 h^2$. Run curve through the spots and you have a complete expansion of the doubling plate. This plate may be ordered in two pieces.

Expansion of a Mast Tube.—In **Fig. 88**, A B C D is the mast tube. Make D E square to D C. Produce centre of the mast and from F as centre describe circle of diameter D E. Divide circumference into say 8 equal parts in the points 1, 2, 3, etc., and produce these parallel to the centre on to lines D B and C A. Extend D E to 1, and A G parallel to D E, produced into the expansion. Girth

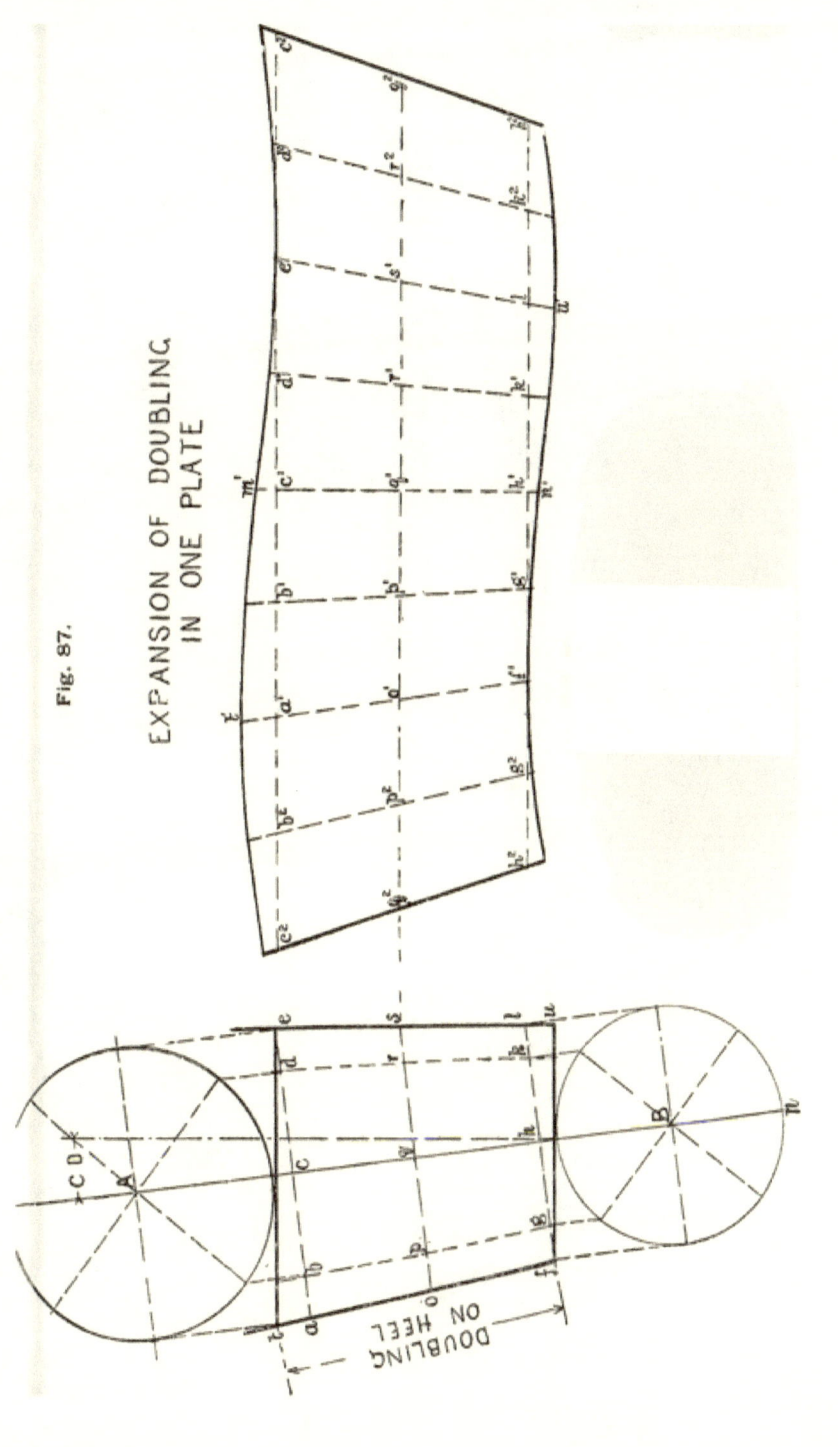

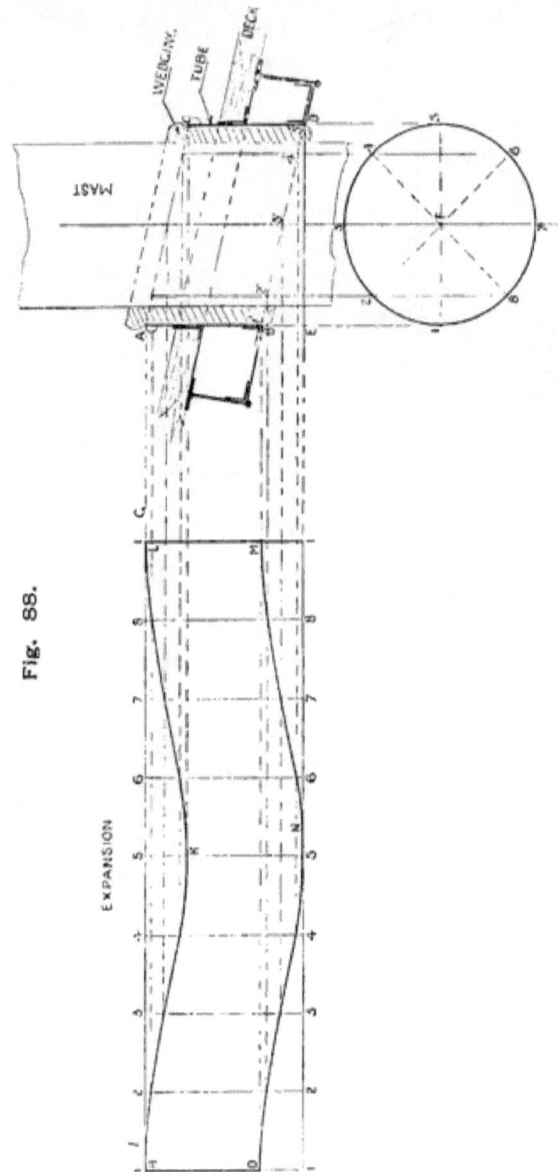

Fig. 88.

circle from 1 for the position of 2, 3, 4, 5, 6, 7, 8 relative to 1, and lay them off in the expansion on D E, produced, from 1. Erect perpendiculars at the points to cut top line A G continued. Produce points 1^1, 2^1, 3^1, 4^1, 5^1, 6^1, 7^1, and 8^1, parallel to D E until they cut their corresponding lines in the expansion. Of course it will be evident that 2^1 and 8^1, 3^1 and 7^1, 4^1 and 6^1 are duplicate and in the same line. Repeat the process for the top of the tube. Then the figure H, K, L, M, N, O, is the developed form of the plate for the tube. It is usual to make these tubes in one plate, and welded at the butt, so that an allowance of about 1 inch should be given for that purpose. The sketch plate may be ordered to prevent waste.

It may be noted that the amount of wedging varies from about $1\frac{1}{2}$ to $2\frac{1}{2}$ inches.

CHAPTER XII.

Rudder Trunk—Expansion of the Trunk—Iron Deck-house—Expansion of the Corner Plate—Cargo Hatch Coamings—Cargo Hatch Coamings with Bell-mouth Bottoms—Marking off the Hawse-pipes—Shaft Tunnel of a Single Screw—Expansion of the Tunnel Plating—Calculation for Expansion—Marking off the Freeboard—Finding the Depth Moulded at the Ship when Dry—Finding the Depth Moulded when the Ship is Afloat—Clipper Stem or Cutwater—End of the Stem Bar or Figure Step—Back of Stem for Lacing-piece—Figure Head Moulds—Forecastle Head—Setting off the Draught Marks on the Stem and Stern—To Form an Oval Manhole—Another Method.

MISCELLANEOUS.

There are several items which are developed either in the drawing-office or on the loft floor, and yet cannot be rightly classed under the term "laying-off." The most important of these we shall now consider in this chapter.

Rudder Trunk.—This trunk is made of steel plate, and should be sufficiently large to allow the rudder to go into position easily, without disturbing the structure. In **Fig. 89** draw rudder in the working position, and then make an outline tracing and place it in the dotted position, so that the top pintle P will clear the gudgeon when shipping, and the stock the deck at B; and draw in the after edge of the trunk C D to suit this. There are two ways of construction, arising out of different methods of fitting the transom floor; where the floor is on the fore side of the stern post, as in **Fig. 91**, the trunk is lapped on to the stern post, but where the floor is on the after side of the post its centre is carried up to the deck sufficiently wide to take the trunk. **Fig. 91** is an enlarged sketch of **Fig. 89**.

Expansion of the Trunk.—In **Fig. 90**, square E to E^1, and extend to E^2; produce $D^1 C^1$ to C^2; make $E^1 C^2$ and $F D^1$ square to $D^1 C^1$. Draw suitable section of bottom, with outstretch $E^1 C^2$; and a section for top, with outstretch $F D^1$. The width will be decided by the size of the attaching bars. Fix points b and c in the bottom section; then girth the section for b, a, and c relative to the floor, and lay the widths out on E^3 and E^2. Through a^1 erect perpendicular for the centre of the expansion. Lift the height of the trunk above $C^2 E^1$ on b^2, and lay it off on the expansion above b^1 and c^1.

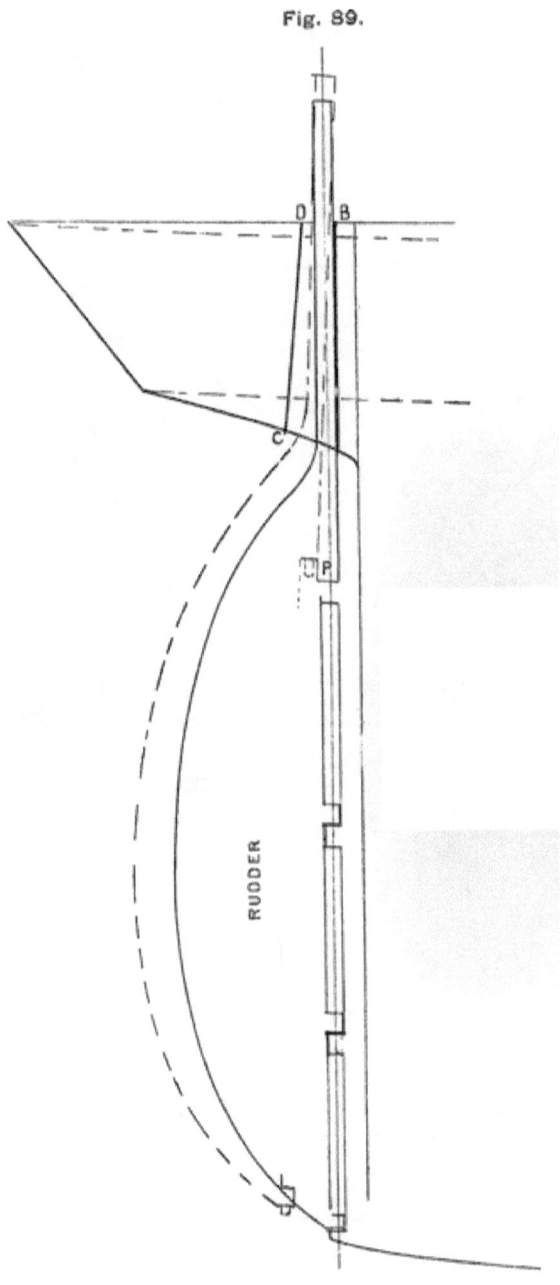

Fig. 89.

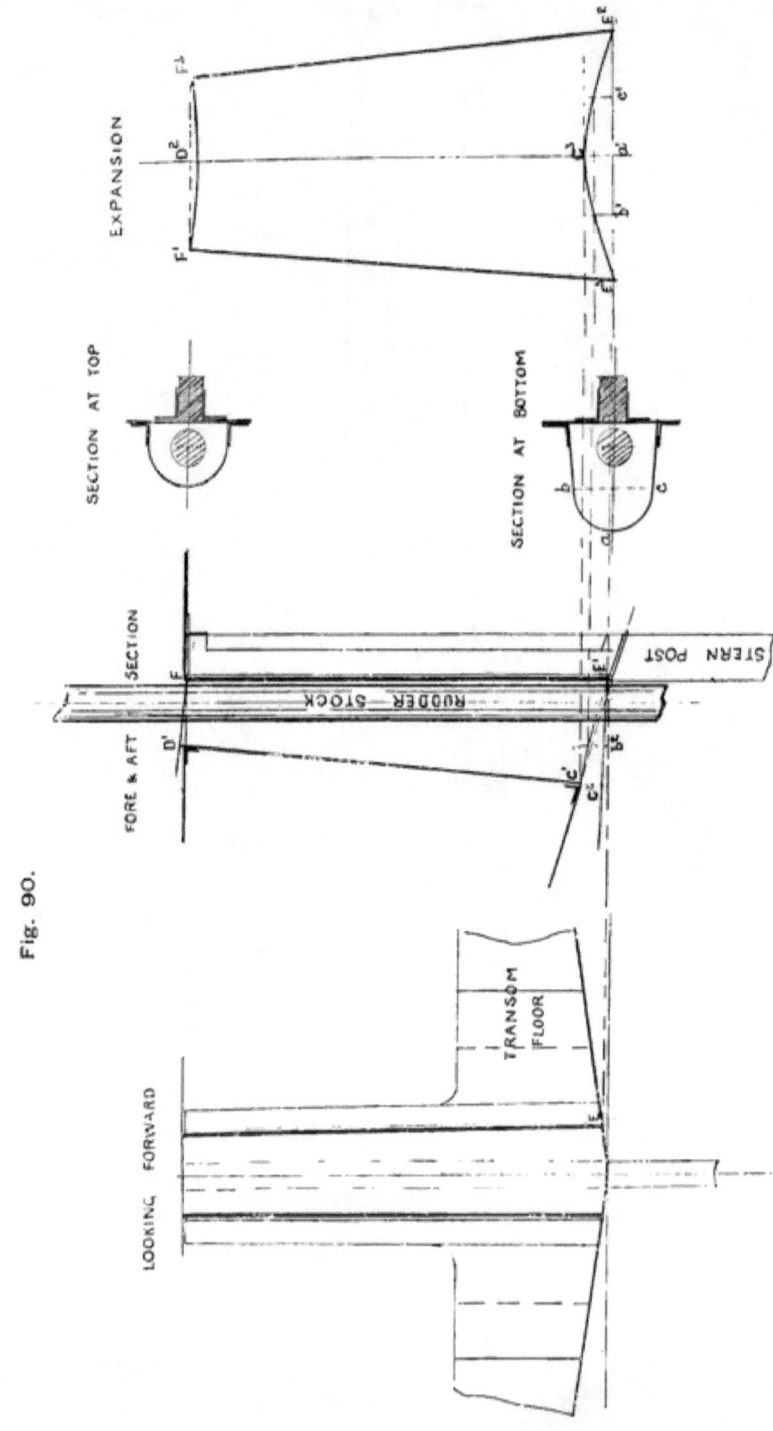

Fig. 90.

MISCELLANEOUS. 99

Measure the length of $C^2 D^1$ with the point C^1 from C^2, and lay them off on the centre above $E^3 E^2$. Pass a curve through the

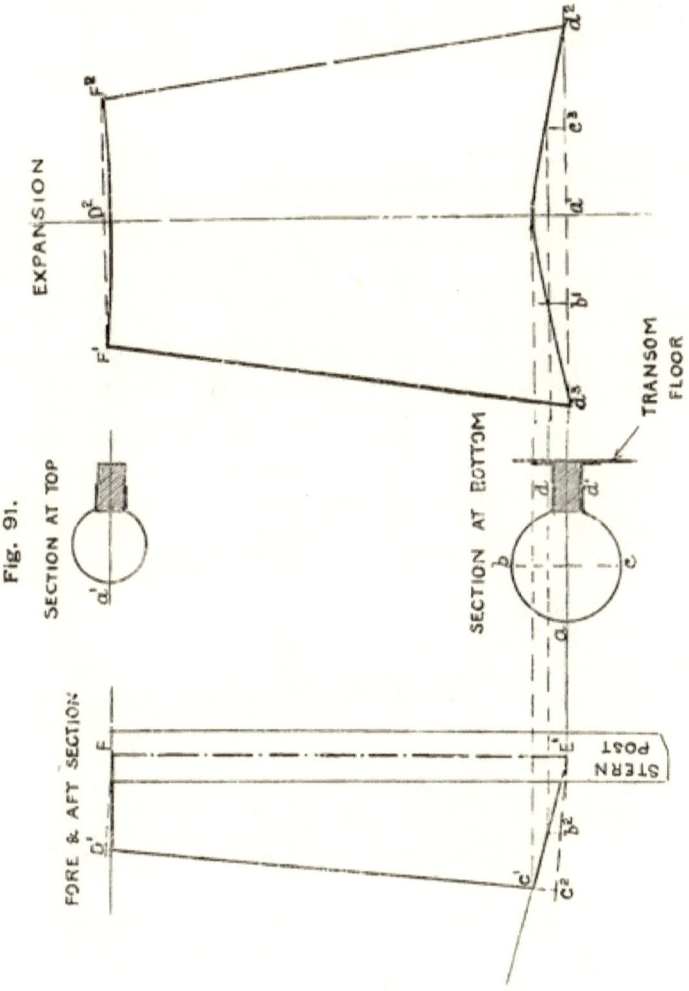

Fig. 91.

points, which will give form of the bottom edge. The point D^2 will be the extreme length of the plate, and the level for the points F^1

will be got by girthing the top section around the centre line, and setting them off each side of D^2. The distance of the trunk top, on the centre below D^2, may be got from D^1 in the fore and aft section, and measuring it down in the expansion. Through the three points run the curve of the top edge.

The expansion of **Fig. 91** is got in a similar manner. Draw circular sections in for top and bottom, with respective outstretch of $F D^1$ and $E^1 C^2$. Mark points a, b, c in the section. Girth the bottom section, and lift points d, b, a, c, and d^1, and lay them out on $d^2 d^2$. Erect perpendicular through a^1. Lift distance above $C^2 E^1$ to the edge of the trunk in fore and aft section, and transfer them to their corresponding points in the expansion for the form of the bottom. Lift C^1 and D^1 above C^2, and set them on the expansion centre above $d^2 d^2$; then D^2 will give level heights for F^1 and F^2. Girth the top section around a^1, and lay the girths on each side of D^2 on $F^1 F^2$. The distance of the centre below D^2 will be got in the fore and aft section below D^1. Join F^1 to d^2, and F^2 to d^2, and you have form of the plate.

Iron Deck-house.—In **Fig. 92**, show side and end elevations with corner plates carried the full depth, and having a radius of about 9 inches. Arrange the coaming plates of suitable height for door step, and place a plate on the top edge about 9 inches deep. These plates are all flush and strapped on the inside. The intermediate plates A are set in and lapped on to corners and top and bottom plates, and made flush at the other butts with pilasters on the outside forming straps. The pilasters are placed to suit doors and windows. This arrangement makes a substantial and good looking job when properly considered. The size of the straight plates may be easily measured off, but in the case of considerable sheer, care must be taken to allow for it in the length of the plates. The ordered length of the corner plates will be C in the figure, while the girth may be calculated. With a radius of 9 inches the width would be

$$\frac{3\cdot1416 \times 18''}{4} + ab + de, \text{ all in inches}, = 14''\cdot14 + 6'' + 6'' = 26''\tfrac{1}{2} \text{ full};$$

a little is allowed for planing.

Expansion of the Corner Plate.—In **Fig. 92**. Erect centre line $g f$, and draw in $h l^1$ and $l h^1$ parallel to the centre and each equal in distance to $\dfrac{14''\cdot14}{2}$ and $m^1 a^1$ and $m n$ parallel and each equal to $\dfrac{26''\cdot14}{2}$ from the centre. Square up d into the elevation, and produce a to

Fig. 92.
DECK HOUSE.

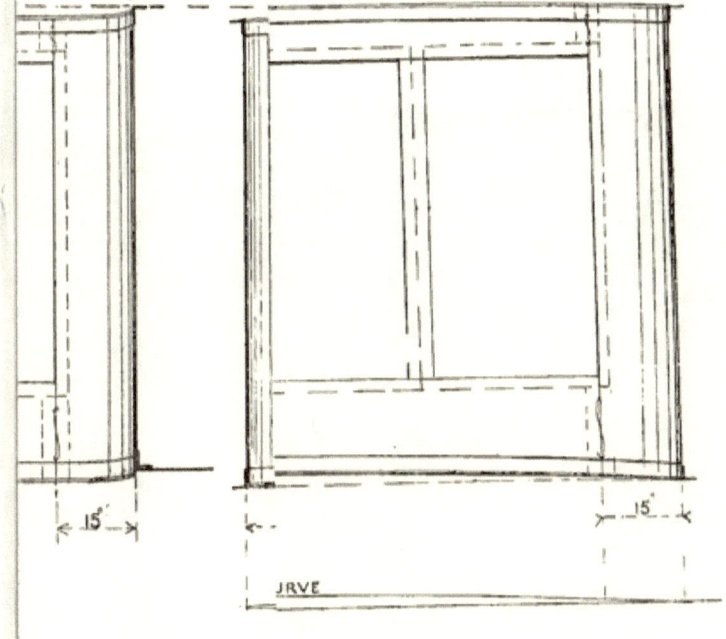

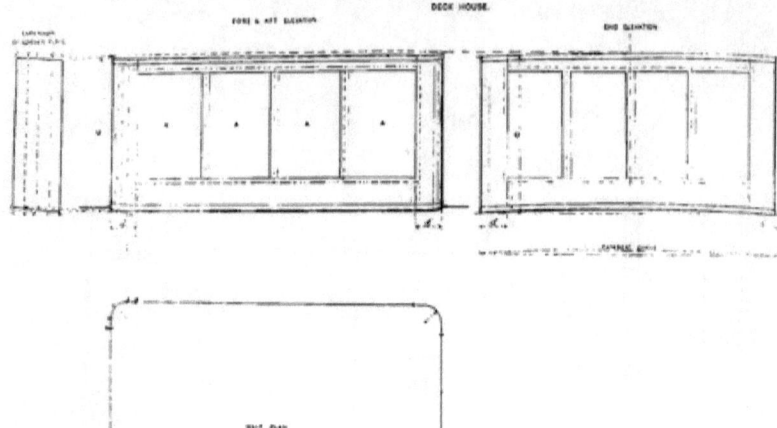

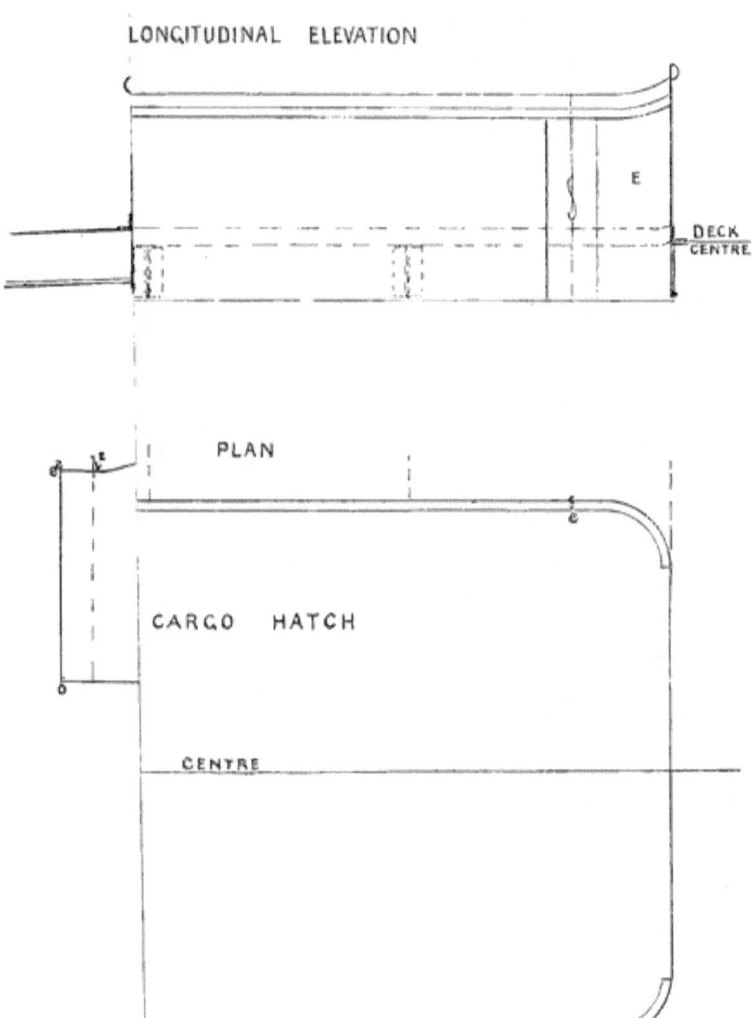

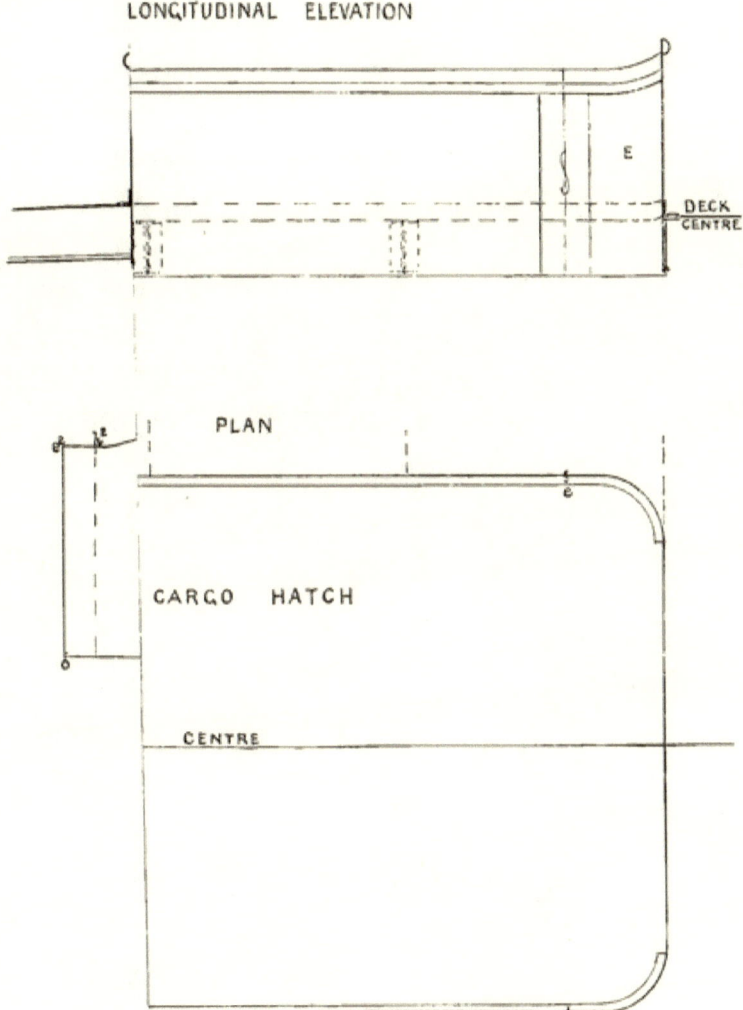

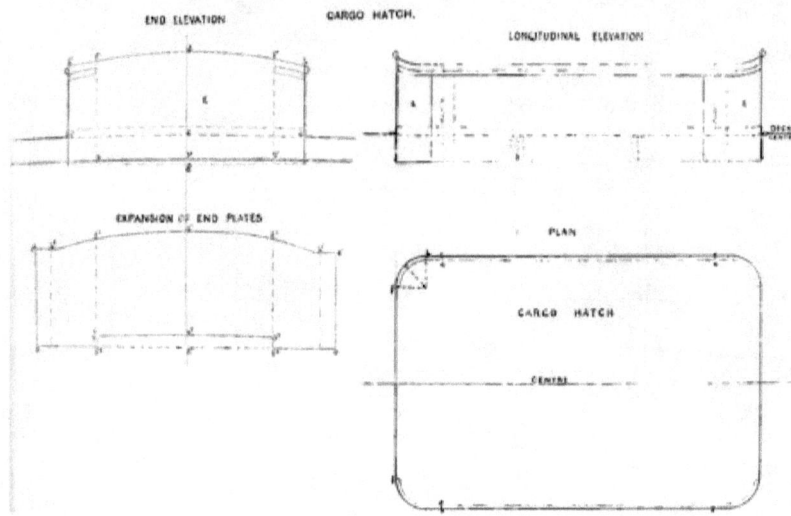

m^1, which will give point on $l h^1$. Then set off a, b, and c on the camber curve and lift the height above the base $o\, m^1$. Set c above $m^1 m$ on gf, and b on $h l^1$, and a on $m^1 n^1$. A curve passed through the points p, w, x, l, m will give the form of the bottom edge of the plate. If the house top has the same camber as the deck then set $m^1 p$ below n, and $w h$ below h^1, and $g x$ below f, and make $u^1 l$ parallel to $l m$. A curve through the spots gives top edge. Then you have expanded form of the plate. Considerable shee. on the deck would make a difference, but if the elevation is rightly drawn down it will be readily seen.

Cargo Hatch Coamings.—In **Fig. 93**, draw the hatch in plan, and arrange butts, c, of the plating, so that the strap will clear beam and corner round. The corners are drawn to about twelve inch radius. Get out transverse form of the deck from the camber curve and set off width of the hatch each side of the centre, from which, above the deck curve, erect perpendiculars, $a^1 b^1$ and $a\, b$, equal to the height at the sides. Run in curve $b^1 d\, b$ with a round up of about 6 inches from the points b and b^1. The sides $a\, b$ and $a^1 b^1$ should extend below the half beams about $\frac{1}{2}$ inch. The end plates E are carried down over $f f$ (in plan) to the top of the bulb of the through beam, and then dropped to meet the side plates at the butts. Next show in longitudinal elevation from the plan and cross section, and square up from the plan the butts c. The size of the side plates may be measured from this plan: should there be any sheer a little must be allowed on the lengths to meet it. The expansion of the end plates E may now be drawn in. Set off the depth $d\, g$ on the centre line of the expansion, $d^1 g^1$, and the form of the top $f^2\, d^1\, f^2 = f^1\, d\, f^1$. Calculate the girth of the corner from f to $h = \dfrac{24'' \times 3\cdot 1416}{4}$ $= 18''\cdot 8496$, say $1'\ 7''$. Measure it off, on each side, parallel to $f^2\, f^3$, shown by dotted line h^1 and h^2; add to it parallel distance $h\, e$ measured from the plan. The depth $e^1\, o$ and $e^2\, o$ is made equal to the depth at the sides, and is carried in parallel to bottom edge over $h^1 e^1$ and $e^2 h^2$, and run in gradually to meet top curve. The bottom edge is recessed between $f^3\, f^3$. The depth at $f^2\, p^3 = f^1\, p$, $d^1\, p^3 = d\, p^1$ and $f^2\, p^2 = f^1\, p^1$. This is owing to end plate over $f f$ stopping at the top edge of the beam bulb. Sketch plates may be ordered from this expansion to form the hatch ends, together with double riveted butt straps. Sometimes the hatch is made broader, across $b^1\, b$ than $a^1\, a$ by about $1\frac{1}{2}$ inches for shipping and unshipping hatch beam. In that case $e^1\, o$ would not be parallel to $d^1\, g^1$.

m^1, which will give point on $l h^1$. Then set off a, b, and c on the camber curve and lift the height above the base $o\,m^1$. Set c above $m^1 m$ on $g f$, and b on $h\,l^1$, and a on $m^1 n^1$. A curve passed through the points p, w, x, l, m will give the form of the bottom edge of the plate. If the house top has the same camber as the deck then set $m^1 p$ below n, and $w h$ below h^1, and $g x$ below f, and make $n^1 l$ parallel to $l m$. A curve through the spots gives top edge. Then you have expanded form of the plate. Considerable sheer on the deck would make a difference, but if the elevation is rightly drawn down it will be readily seen.

Cargo Hatch Coamings.—In **Fig. 93**, draw the hatch in plan, and arrange butts, e, of the plating, so that the strap will clear beam and corner round. The corners are drawn to about twelve inch radius. Get out transverse form of the deck from the camber curve and set off width of the hatch each side of the centre, from which, above the deck curve, erect perpendiculars, $a^1 b^1$ and $a b$, equal to the height at the sides. Run in curve $b^1 d b$ with a round up of about 6 inches from the points b and b^1. The sides $a b$ and $a^1 b^1$ should extend below the half beams about $\frac{1}{2}$ inch. The end plates E are carried down over $f f$ (in plan) to the top of the bulb of the through beam, and then dropped to meet the side plates at the butts. Next show in longitudinal elevation from the plan and cross section, and square up from the plan the butts e. The size of the side plates may be measured from this plan: should there be any sheer a little must be allowed on the lengths to meet it. The expansion of the end plates E may now be drawn in. Set off the depth $d y$ on the centre line of the expansion, $d^1 g^1$, and the form of the top $f^2 d^1 f^2 = f^1 d f^1$. Calculate the girth of the corner from f to $h = \dfrac{24'' \times 3\cdot 1416}{4}$ $= 18''\cdot 8496$, say $1'\ 7''$. Measure it off, on each side, parallel to $f^2 f^2$, shown by dotted line h^1 and h^2; add to it parallel distance $h\,e$ measured from the plan. The depth $e^1 o$ and $e^2 o$ is made equal to the depth at the sides, and is carried in parallel to bottom edge over $h^1 e^1$ and $e^2 h^2$, and run in gradually to meet top curve. The bottom edge is recessed between $f^3 f^3$. The depth at $f^2 p^2 = f^1 p$, $d^1 p^5 = d\,p^1$ and $f^2 p^2 = f^1 p^1$. This is owing to end plate over $f f$ stopping at the top edge of the beam bulb. Sketch plates may be ordered from this expansion to form the hatch ends, together with double riveted butt straps. Sometimes the hatch is made broader, across $b^1 b$ than $a^1 a$ by about $1\frac{1}{2}$ inches for shipping and unshipping hatch beam. In that case $e^1 o$ would not be parallel to $d^1 g^1$.

Cargo Hatch Coamings with Bell-mouth Bottoms.—In **Fig. 94**, sketch in the hatch as before, only with suitable bell mouth on the lower edge as shown in the plan by the dotted lines a, b, c, and d. The corner plates are butted on the ends rm, and on sides sn, to allow them to be easily "blocked" into shape. The expansion of the end plates may be got by girthing ef, and setting it off on the cross section $f^1 e^1$, the bottom line being made parallel to gh. Then $he^1gf^1r^1$ is the expanded plate. The side plates are got by girthing kr and setting it off in the longitudinal elevation $k^1 l^1$ and $k^2 l^2$, then $k^1 l^1 l^2 k^2$ is the expansion. The corner plates will need more careful attention. Girth the length on fe, from the point r^1 squared over, for t and e, and lay them out on the expansion centre line from $t^1 s^1$. Square down t into the plan, and draw in through the point quadrant of a circle op. Then girth, in the plan, around the centre of the corner mn, op, rs, and set these on their corresponding lines in the expansion $m^1 n^1$, $o^1 p^1$, $r^2 s^1$. Produce the heights from the cross sections to meet and you have spots for the form of the expanded plate. A few extra points between rs would enable you to get the top edge more accurately.

Marking off the Hawse-pipes.—Indicate on the forecastle deck the lead of the cable from the windlass, and settle the position of the centre of the pipe on the forecastle deck to clear beams, through which make a hole about 6 inches square and drop a plumb-line from it. Fix the position of the centre of the pipe on the shell, by continuing the plumb-line distance parallel to the centre of the ship, and in such a position that the stringer bar remains undisturbed and only one frame to cut. Cut a hole at this point in the shell about 4 inches square. Obtain a straight round bar, about $1\frac{1}{2}$ inches in diameter, long enough to reach from the forecastle hole to that of the shell, and secure it by wedges in position at the proper distance from the centre line. Make a piece of wood, say 2 inches thick, of the form A, shown in **Fig. 95**, and of radius D equal to half the diameter of the outside of the pipe. Hollow one edge of the wood to fit the bar, chalk the point C and slide it round the bar, so that C will touch the shell. By this means the correct shape of the hole required will be secured. Reverse the piece A, and do the same with the forecastle deck. Take the bar out of position and cut the holes; which should be made slightly larger for the easy shipment of the pipe. A circular pipe mould, sufficiently long to reach beyond the holes, is then placed in position; that of the

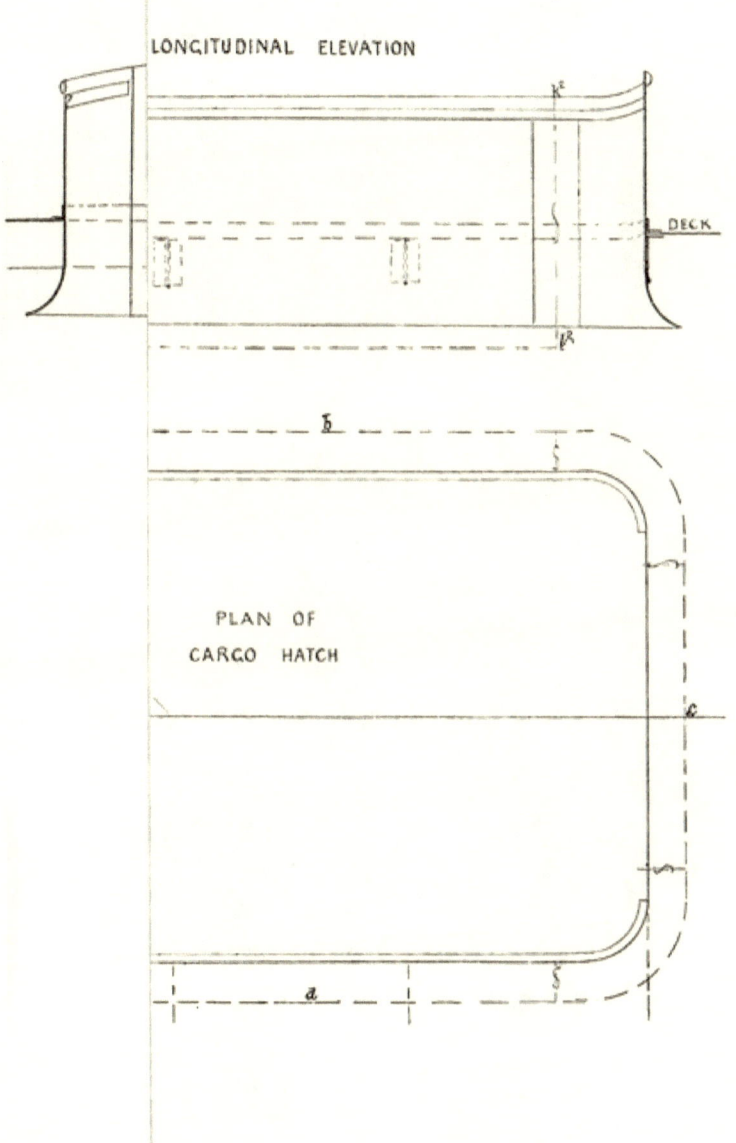

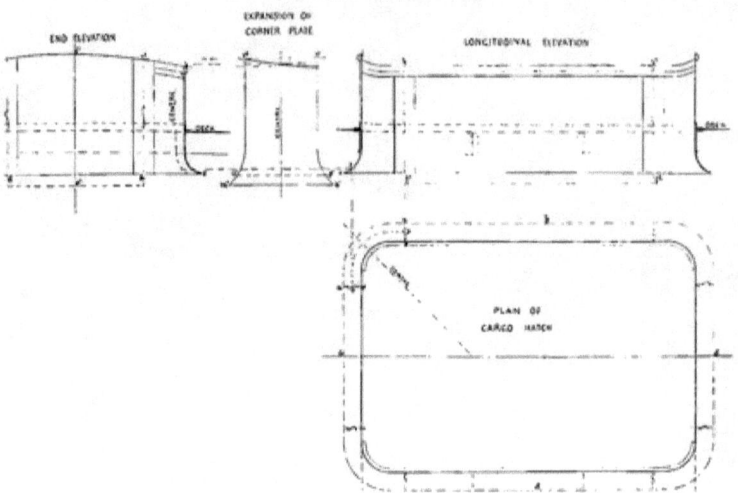

Fig. 94.
BELL-MOUTH CARGO HATCH.

shell is cut flush and a ring mould for shell flange screwed on to the end. It may be noted that this mould is shipped into position from the outside of the vessel, and has a slight taper on the end put in first. When in position a thin layer of lead is tacked on the forecastle deck end, to show the line of the deck all round, which is the correct length; it is then drawn out and sent away to the foundry for casting the pipe.

Shaft Tunnel of a Single Screw.—The shaft is cased in from the engine room to the after stuffing-box bulkhead, forming a tunnel; so that it is accessible to the engineers at all times. Get out a plan and

Fig. 95.

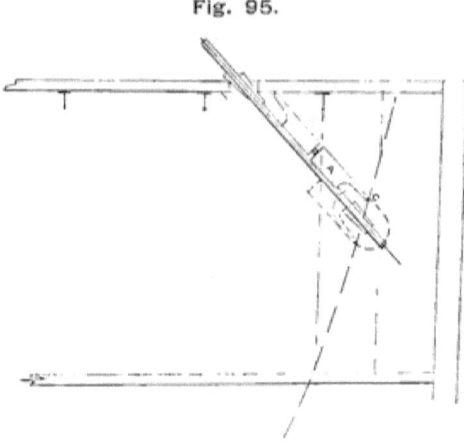

elevation on ¼ inch scale as in **Fig. 96**, showing the floors and tank top between the bulkheads. Set off at each bulkhead, in the elevation, the height of the centre of shaft and draw line in. Arrange thrust recess at the fore end, in plan and elevation, usually of box form. The tunnel crown just forward of the after bulkhead is attached to the side stringers, and an additional partial bulkhead is fitted, G H, extending to ship's side, against which the form from C to D terminates, and enables a better connection to be made, and leaves more room to get at the stuffing-box. Set off height of tunnel at the ends F C and E D, and connect C to D. Then draw in plan the sides of the tunnel from C to D. The passage way is shown on the starboard side of the centre. Arrange the stiffeners; which may be on the out or inside—they are shown inside. The spacing is 4 feet, but

shell is cut flush and a ring mould for shell flange screwed on to the end. It may be noted that this mould is shipped into position from the outside of the vessel, and has a slight taper on the end put in first. When in position a thin layer of lead is tacked on the forecastle deck end, to show the line of the deck all round, which is the correct length: it is then drawn out and sent away to the foundry for casting the pipe.

Shaft Tunnel of a Single Screw.—The shaft is cased in from the engine room to the after stuffing-box bulkhead, forming a tunnel; so that it is accessible to the engineers at all times. Get out a plan and

Fig. 95.

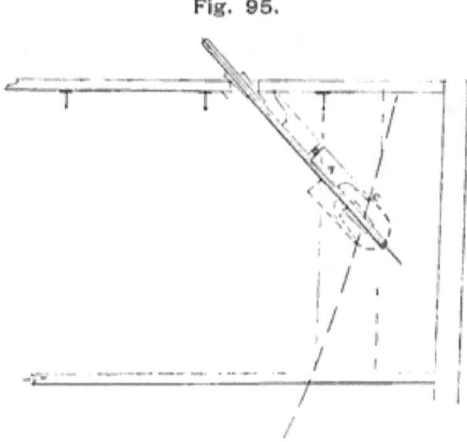

elevation on ¼ inch scale as in **Fig. 96**, showing the floors and tank top between the bulkheads. Set off at each bulkhead, in the elevation, the height of the centre of shaft and draw line in. Arrange thrust recess at the fore end, in plan and elevation, usually of box form. The tunnel crown just forward of the after bulkhead is attached to the side stringers, and an additional partial bulkhead is fitted, G H, extending to ship's side, against which the form from C to D terminates, and enables a better connection to be made, and leaves more room to get at the stuffing-box. Set off height of tunnel at the ends F C and E D, and connect C to D. Then draw in plan the sides of the tunnel from C to D. The passage way is shown on the starboard side of the centre. Arrange the stiffeners; which may be on the out or inside—they are shown inside. The spacing is 4 feet, but in way of the cargo hatches they are required to be 3 feet. Those on forward recess are the same spacing as the ship's framing. Show stools, etc., in position.

Expansion of the Tunnel Plating.—The forward recess is easily marked off. Get out a section through A B, arrange laps, flanging the plates as shown at the sides. Girth this section, around the centre of the tunnel, at the forward and after ends, and set off the widths in the expansion at each side of the centre, and draw lines in.

The part from C to D needs more care. Calculate the circumference of a circle of diameter equal to the width of the tunnel, that is, $2a$, and half what you get will be the girth of the circular part; then add on for the after end sides $cb + fd$, which is the total girth without laps; the forward end will be found by adding to the circular girth $cg + fe$. These total dimensions are laid off at their respective ends in equal halves, at each side of the centre of tunnel, which will give the expansion of form from C to D.

A lap is placed at the centre of the tunnel as shown, and the others are arranged so that the two side plates do not require any round. In other words laps marked o and p in the section are kept below e and f.

Calculation for Expansion Width at C F and E D.—Height at C F = 7 feet 9 inches, E D = 7 feet 3 inches. Tunnel 4 feet wide. Circumference of circular part to $ef = \dfrac{4 \text{ feet} \times 3\cdot1416}{2} = \dfrac{12\cdot5664}{2} =$ 6·2832 feet = 6 feet $3\frac{3}{8}$ inches.

Then girth at C F.		Then girth at E D.	
Circular part	= 6' $3\frac{3}{8}$"	Circular part	= 6' $3\frac{3}{8}$"
Lower part 2 (7' 9" −		Lower part 2 (7' 3" −	
2' 0") = 5' 9" × 2	= 11' 6"	2' 0") = 5' 3" × 2	= 10' 6"
Laps 3 × 2$\frac{1}{2}$"	= 7$\frac{1}{2}$"	Laps 3 × 2$\frac{1}{2}$"	= 7$\frac{1}{2}$"
Total girth of plating	18' 1$\frac{7}{8}$"	Total girth of plating	17' 4$\frac{7}{8}$"

Marking off the Freeboard.—**Fig. 97** is the marking for a steamer, and **Fig. 98** for a sailing ship. The disc and lines must be permanently indicated by centre punch marks or cutting. The centre of the disc is required to be placed midway between the perpendiculars of the load-line on each side of the vessel. It is difficult to exactly state a method for marking it off on the ship, for ships vary so much in their construction. One method is to place a deal, levelled, on top of the sheer strake (see **Fig. 99**), and drop a plumb-line from it.

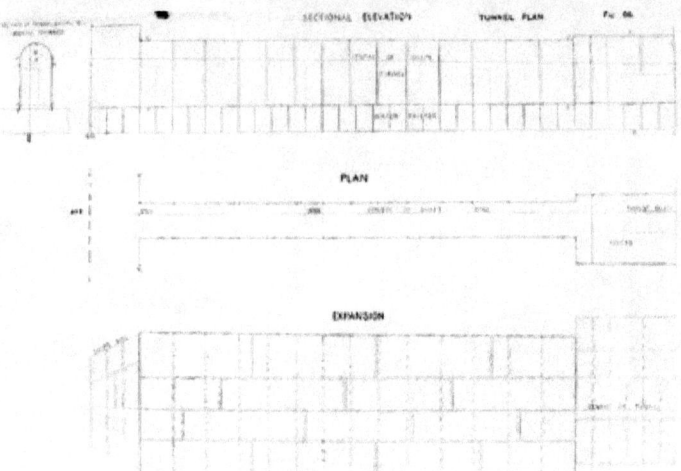

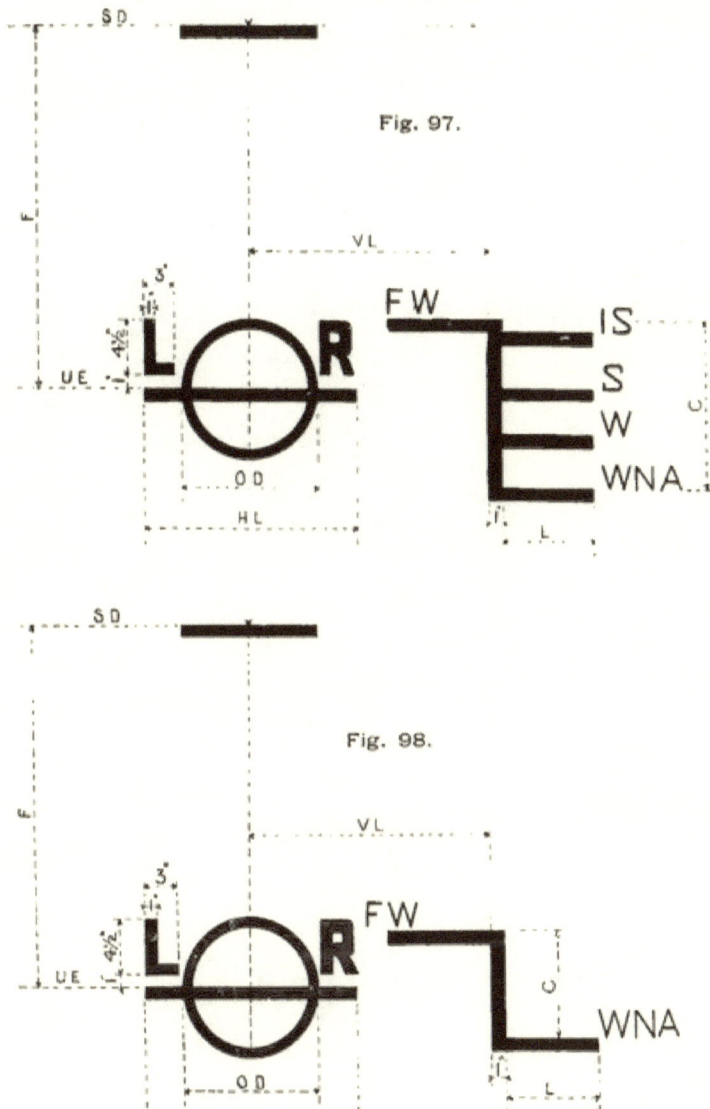

Fig. 97.

Fig. 98.

S D = Top of Statutory Deck Line.
F   = Freeboard.
U E = Upper Edge of Horizontal Line passing through Centre of Disc.
V L = 21 Inches.
O D = Outside Diameter of Disc, 12 Inches, 1 Inch Thick.
H L = Horizontal Line 18 Inches Long and 1 Inch in Thickness.
L   = 9 Inches Long and 1 Inch Thick.
C   = These Dimensions to be taken from Centre of Disc to Top of each Line.

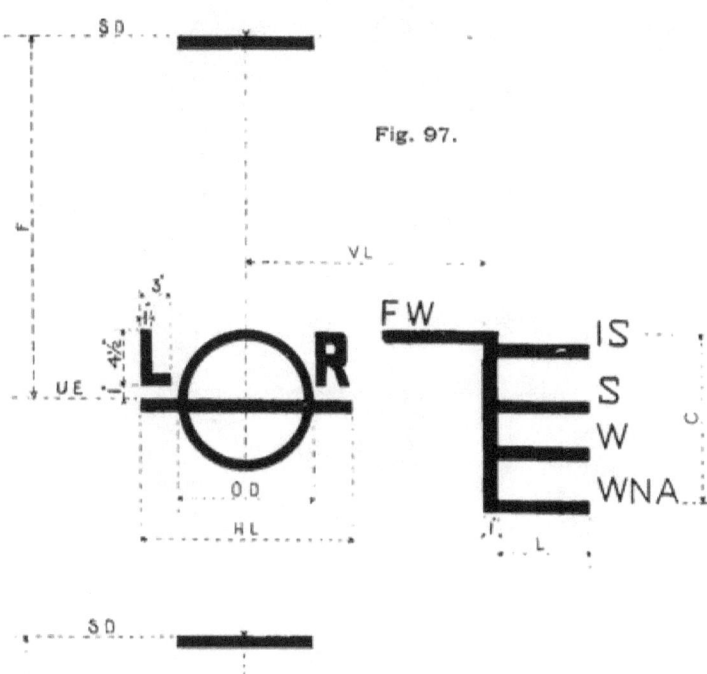

Fig. 97.

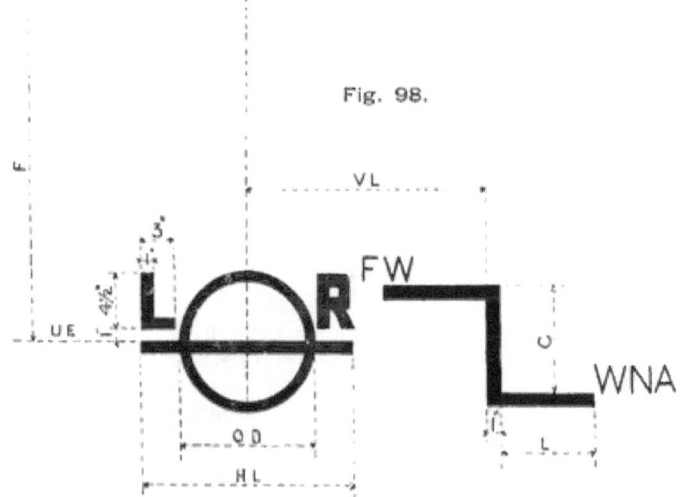

Fig. 98.

S D = Top of Statutory Deck Line.
F = Freeboard.
U E = Upper Edge of Horizontal Line passing through Centre of Disc.
V L = 21 Inches.
O D = Outside Diameter of Disc, 12 Inches, 1 Inch Thick.
H L = Horizontal Line 18 Inches Long and 1 Inch in Thickness.
L = 9 Inches Long and 1 Inch Thick.
C = These Dimensions to be taken from Centre of Disc to Top of each Line.

Mark on side plating the depth moulded line B, and above it indicate the statutory deck line S D. Measure the height from S D to the deal and add it to the required freeboard. Mark this on a straight batten, and set it below the deal on the plumb-line, and square the point on to the shell, which will give the centre of the disc.

Finding the Depth Moulded at the Ship when dry.—This is done when the vessel is on the yard, or in dry dock, by fixing a deal

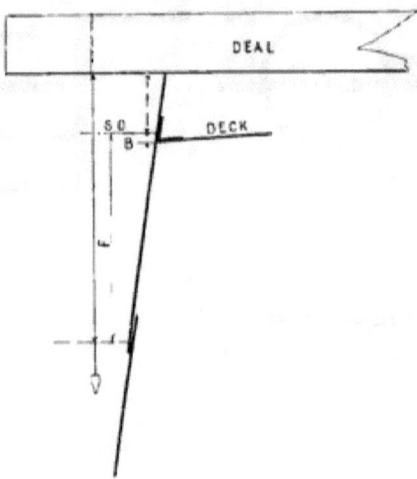

or straight edge as A B in **Fig. 100,** under the keel athwartships in way of the midship frame, and one above it on the top of the sheer strake edges C D. Drop a plumbed tape line from the top deal, clear of the side, and measure the depth between the inside edges of the deals on each side of the vessel. Deduct the height of the sheer strake above the beams and the depth of the keel from each, and add the remaining quantities together, then half the addition will be the depth moulded.

Finding the Depth Moulded when the Ship is afloat.— Fix a deal across the ship at the midship frame on the top of the sheer strakes or rail. Measure the plumbed distance from the under side of the deal to the water on each side of the vessel, and deduct the height of the sheer strake from the top of the beams in each case. To

808

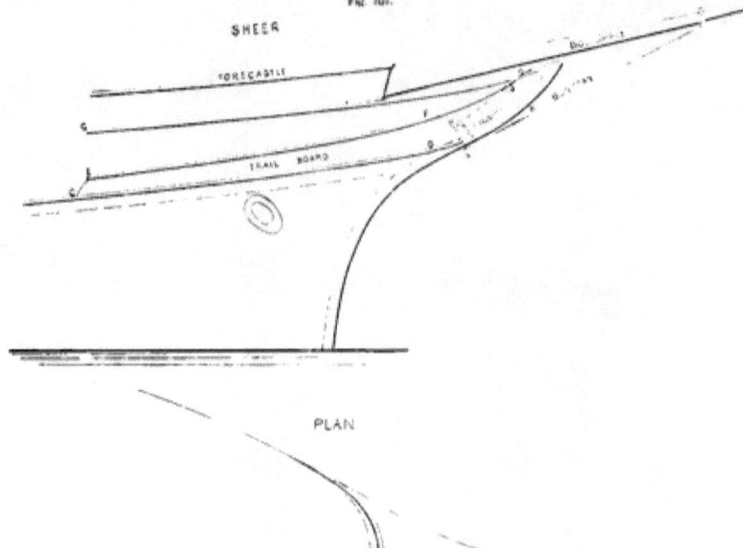

Fig. 101.

SHEER

FORECASTLE

TRAIL BOARD

PLAN

the remaining quantities add the mean draught, less the depth of the keel, then half the sum of these will be the depth moulded.*

Clipper Stem or Cut-water.—The arrangement is very much a subject of taste, etc., but **Fig. 101** is given as near as possible on the lines of modern ideas. Draw in the lower moulding C D of the trailboard, which is a continuation of the gunwale moulding, and made of hard wood. It should terminate on the foot of the figure about the

Fig. 100.

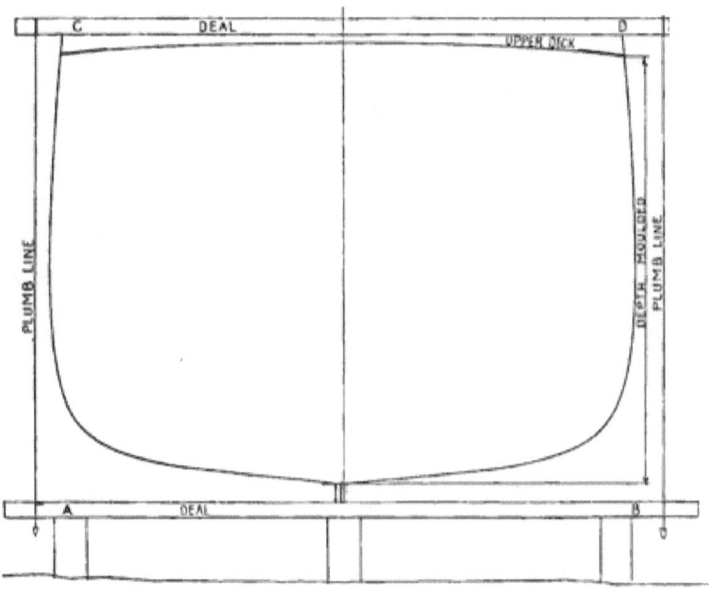

step A B. Run in the top moulding E F, lifting it up to form the crown of the figure and cutting the rail line G H at H, produced to the correct sheer. Then draw in the profile of the figure. The bowsprit should be shown in position to clear the figure by about 2 inches. The forecastle head is usually made a continuation of the part of the stem just above the load-line.

End of the Stem Bar or Figure Step.—The bobstay eye is put on the end of the stem bar, which should terminate in such a

* The mean draught is half the sum of the forward and after draughts.

the remaining quantities add the mean draught, less the depth of the keel, then half the sum of these will be the depth moulded.*

Clipper Stem or Cut-water.—The arrangement is very much a subject of taste, etc., but **Fig. 101** is given as near as possible on the lines of modern ideas. Draw in the lower moulding C D of the trailboard, which is a continuation of the gunwale moulding, and made of hard wood. It should terminate on the foot of the figure about the

Fig. 100.

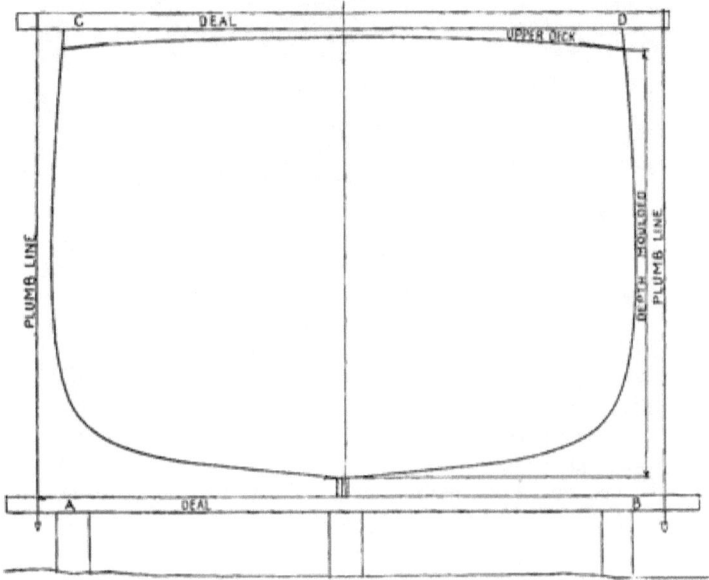

step A B. Run in the top moulding E F, lifting it up to form the crown of the figure and cutting the rail line G H at H, produced to the correct sheer Then draw in the profile of the figure The bowsprit should be shown in position to clear the figure by about 2 inches. The forecastle head is usually made a continuation of the part of the stem just above the load-line.

End of the Stem Bar or Figure Step.—The bobstay eye is put on the end of the stem bar, which should terminate in such a

* The mean draught is half the sum of the forward and after draughts.

position that the stay when fixed will clear the front, and yet allow sufficient for forming figure of 6 to 8 feet long. This will decide the step A B, which is made square to the cut-water and extends back to B, about 9 inches clear of the stem bar.

Back of Stem for Lacing-piece.—From the point B, in Fig. 104, draw B H a straight line to give about 22 inches across the figure at J K. The side plating projects forward of B H about 2 inches, as shown in sectional **Fig. 102**, and the distance $a b$ should be about 8 inches for the full depth of B H, or slightly tapered towards B. A plate is fitted to form back, being flanged to take side plating and bent round at foot to form the step A B, upon which is fitted a shoe, as in **Fig. 103**, which takes the shell and is attached with a tap to the stem head, and serves for stepping the figure foot into. A lacing-piece of $1\frac{1}{2}$ inch pine is attached to the back plate B H for the full depth. The figure is fixed by passing bolts, from the front, through the back plate with nuts on the inside.

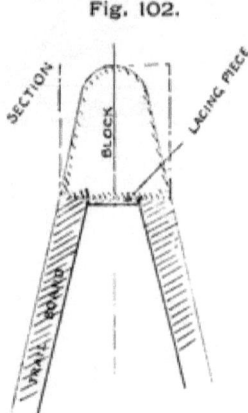

Fig. 102.

Figure Head Moulds.—The form of the trail board is painted on the ship to please the eye, and a mould made therefrom embracing top rail, which is sent to the carver, together with one for the depth and length of the block, and one for the shape of the lacing-piece and step.

The trail-boards are made about 4 inches thick to begin with, so that the figure-block should be about 16 inches wide. This is cut down in thickness in the carving.

Forecastle Head.—The top, between the rails, forward of forecastle head, is plated over and attached at the sides by an angle bar worked on the underside, or the plating is flanged down on to the side plating. The forecastle head is connected to the top plating by an angle bar fitted on the topside with packing iron in the feather edge.

Setting off the Draught Marks on the Stem and Stern.—Fig. 105 shows the outline of a vessel on the stocks. The bottom of the keel should be a straight line. It is tested by placing "dark-sight"

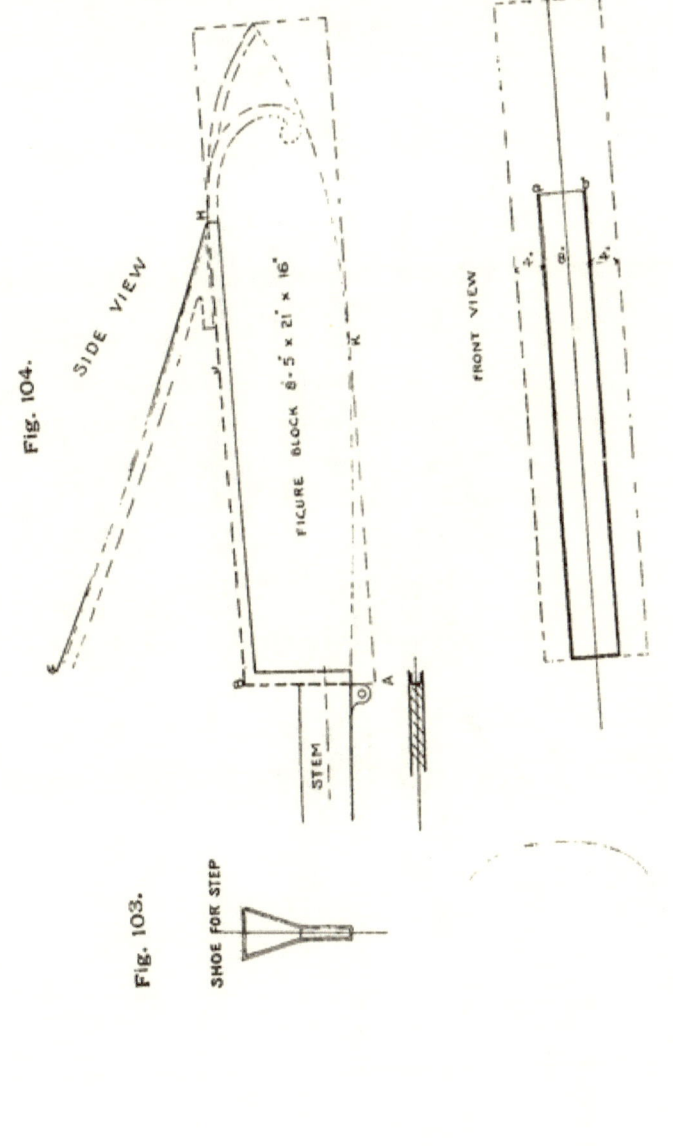

Fig. 105.

LONGITUDINAL ELEVATION

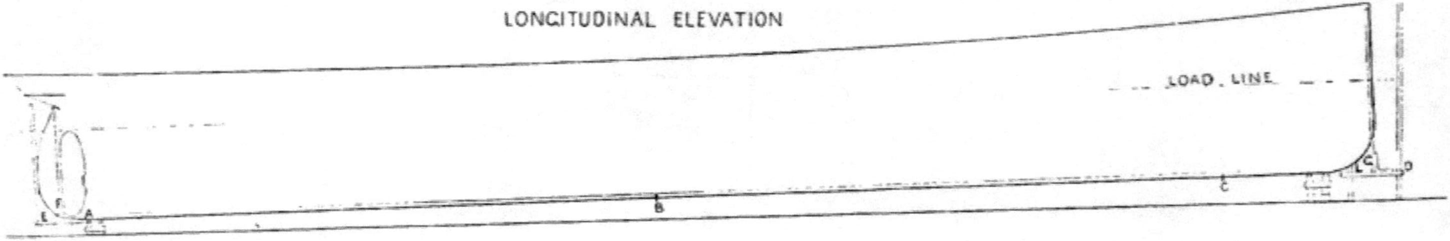

LOAD LINE

boards, like H in **Fig. 106**, with an open space P of about ¼ of an inch, on a level with the bottom of the keel at various positions in the length of the vessel, A, B, C; and adjusting A and C until the sight holes are in a line with B and each other, and so that the sight edge embraces the extreme bottom points of the keel. The sight line may then be produced on to D—any upright at the stem—and a temporary shore placed in for the purpose will secure L in the same way. To these points L and D a straight edge J is attached. Then stand the measuring batten G, as shown, on top of this, keeping it square to the

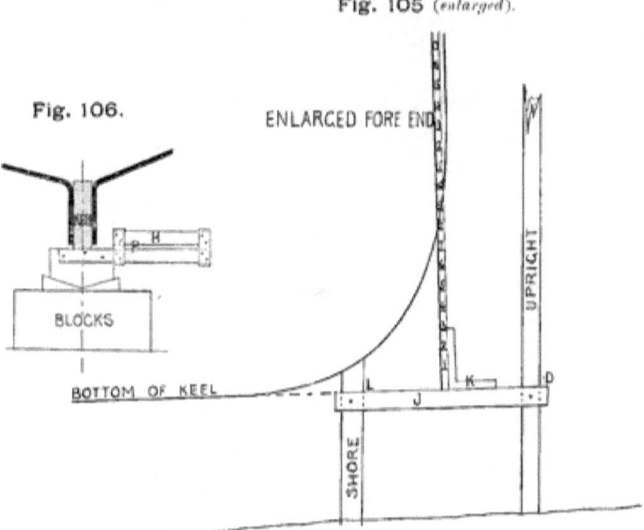

Fig. 105 (*enlarged*).

Fig. 106.

edge L D with set square K, and line off the foot marks on the stem to about 3 feet above the load-line. It is not necessary to place any figures, except in shallow draught boats, below about 4 feet, as the ship is never likely to draw less water. The after end may be lined off in the same manner by producing C B A to E and placing straight edge to points E A. Measure up F from A E the draught marks, and stencil the figures in; after which the driller should cut them in with a chisel, used expressly for that purpose. The figures are 6 inches deep, and are set off as on the batten G.

Fig. 107.

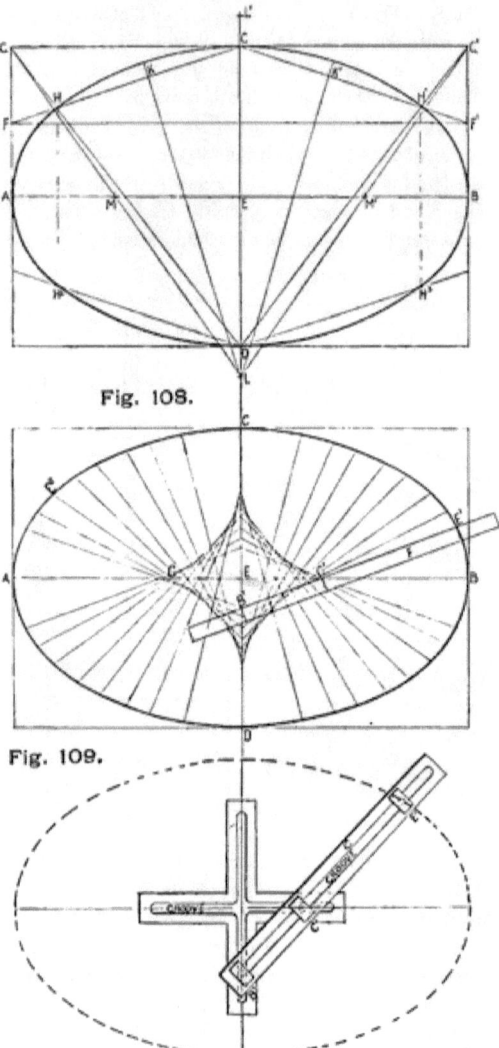

Fig. 108.

Fig. 109.

To Form an Oval Manhole.—In **Fig. 107**, let A B be the longest dimension. Bisect it in E, and erect perpendicular of indefinite length each side of A B. Set off E C and E D each equal to half the shortest dimension. Draw A G and B G¹ parallel and equal to E C. Bisect A G in F and B G¹ in F¹. Join F and F¹ to C. Bisect E A and E B in M and M¹ and join G to M and produce to L, G¹ to M¹ and produce to L, cutting C F and C F¹ at H and H¹. Then bisect C H and C H¹ in K and K¹. From K and K¹ drop perpendiculars cutting centre line C L, and from the cutting point draw arc H C H¹. Then from M and M¹ as centres draw arcs H² A H and H³ B H¹ to meet H C H¹. Complete the opposite side in the same manner.

Another Method.—In **Fig. 108**, draw A B and C D, bisected by A B in the point E and at right angles. On a slip of paper F mark the length E¹ B¹ = E B, and from E¹ make E¹ C¹ to equal E C. Work the slip into any position so that B¹ is on C D, and C¹ on A B, then E¹ will give on the line B¹ C¹ produced a point for the curve. It will be readily seen that any amount of points may be found for drawing in the oval. They are shown by lines across the figure. An easier method still is to construct an instrument, **Fig. 109**, with grooves for the movable arm C to work in, with the points B¹ and C¹ fixed by thumb screws, and the point E¹ fixed, but having a pencil on the lower side for drawing in the oval.

Different sized ovals are drawn, and an iron template made of each, with a hand-hold for the use of the yard in marking off.

WAR VESSELS.

CHAPTER XIII.

Protective Deck in a Cruiser: To Obtain the Form—Fairing-up the Form—Ordinary Expansion—More Correct Method of Expansion—Mode of Plating—Model of Deck—Bevels for Beams. *Belt Armour and Deck in a Battle Ship:* General Description of the Structure—Correction on Loft Floor for Belt Armour in Fairing-up the Moulded Form—Belt Deck—Its Support—Connection—Butts and Seams—Belt Armour on Box-Framing—Armour Shelf—Protective Deck at Ends in a Battle Ship—Finish of Belt Armour at the Ends—Fairing-up the Belt Armour—Expansion—Moulds Required for Ordering Plates. *General Description of the Structure of Barbettes or Redoubts:* Expansion of the Armour—Ordering the Armour—Moulds Required—Expansion of Inner Thick Plates—Circular Barbette—Revolving Turret in Redoubt.

The principle of "laying-off" the moulded form, on the loft floor, of this class of vessel is the same as in merchant ships, so that it would be useless to travel over the ground again. While this is the case, there are special features connected with the armour, shell plating, double bottoms, gun galleries, moulds, and other considerations in composite and sheathed vessels, which do not enter into iron or steel merchant vessels; and, therefore, need some explanation here. It is proposed to devote a few chapters to these principal points.

ARMOUR.

Protective or Armour Deck.—Almost all war vessels, where side armour is not fitted, have a protective or armour deck extending the entire length, or at least over the machinery space, formed of two or three layers of thick plates. It is placed about the load-line. In some cases it is of such a simple character that it needs no explanation, except that given for ordinary decks on page 31; but where the modified turtle back form is adopted the following method may be found useful.

To Obtain the Form of the Protective Deck.—In the Body plan, **Fig. 110**, arrange on the midship section the distance below the load-line, and from this point A the angle of the deck to the ship's side A T. The centre height T U, above the load-line, should be

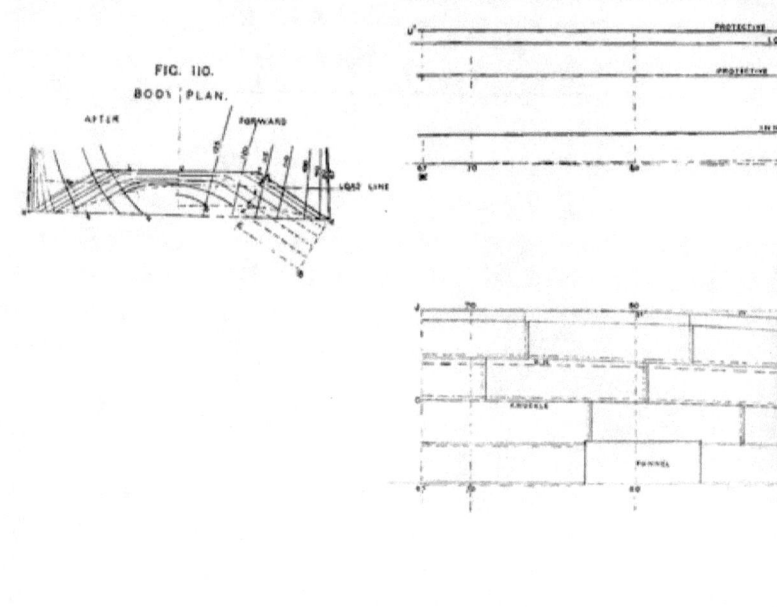

FIG. 110.
BODY PLAN.

sufficiently high to get machinery comfortably into position. Level this point U out into each Body until it cuts the produced side line A T, and reduce the knuckle T by a small curve, or radius. Place in the Sheer **Fig. 111** the deck at the centre U, which is generally maintained for the length of the machinery space. In some cases it is dropped sharply abaft and forward of machinery bulkheads, to get 'tween deck height for the crew space, where it is made almost flat, with a short curve down to meet the side line. Some have it considerably heightened amidships to suit engines, etc. In **Fig. 111**, it is drawn a fair continuous curve to the after and forward extremities. At the side it is kept, as near as possible, the same distance below the load-line over the machinery and magazine spaces, and then gradually faired in to the end points. At the after end it covers the rudder head, etc., while at the fore end it is attached to a spur springing from the ram (see **Fig. 112**). After these initial lines are arranged in the Sheer, the centre heights and the side, relative to the load-line, are lifted on the frame stations and transferred into the Body. Level lines are drawn through the centre points towards the sides, and the side lines produced at about the midship angle, or parallel to A T to meet the knuckle, reduced by a small curve. At the extreme ends it takes an ordinary camber form. In this way the form, on a few sections, is approximated. It is necessary, to prevent twist on the plates, that the sections be made parallel to A T.

Fairing-up the Form of the Protective Deck.—This may be done on the buttocks, but it is best to use diagonal lines like D E and F G. The cutting points of the trial form are lifted from the level line A H, and faired in the Sheer on the contracted method. After a few sections are faired, the intermediates can be easily run in by obtaining spots on the diagonals and heights for the centre from the Sheer, **Fig. 111**.

Expansion of the Protective Deck.—Assuming that the deck is a continuous fair surface, the general method is to girth the centre line $U^1 Y^1$ in **Fig. 113**—the Sheer enlarged for the fore end—for the position of the frame stations, stern, and stem terminations. Set these along a straight line representing the centre line of the vessel, and erect perpendiculars from the points, which are the expanded frame stations. On these lay off the girthed width, on corresponding sections, of the deck from the centre line : a curve through the spots will give J W X Y in **Fig. 114**. Repeat the process for the other side of the vessel. Upon this expansion all the openings, longitudinal

sufficiently high to get machinery comfortably into position. Level this point U out into each Body until it cuts the produced side line A T, and reduce the knuckle T by a small curve, or radius. Place in the Sheer **Fig. 111** the deck at the centre U, which is generally maintained for the length of the machinery space. In some cases it is dropped sharply abaft and forward of machinery bulkheads, to get 'tween deck height for the crew space, where it is made almost flat, with a short curve down to meet the side line. Some have it considerably heightened amidships to suit engines, etc. In **Fig. 111**, it is drawn a fair continuous curve to the after and forward extremities. At the side it is kept, as near as possible, the same distance below the load-line over the machinery and magazine spaces, and then gradually faired in to the end points. At the after end it covers the rudder head, etc., while at the fore end it is attached to a spur springing from the ram (see **Fig. 112**). After these initial lines are arranged in the Sheer, the centre heights and the side, relative to the load-line, are lifted on the frame stations and transferred into the Body. Level lines are drawn through the centre points towards the sides, and the side lines produced at about the midship angle, or parallel to A T to meet the knuckle, reduced by a small curve. At the extreme ends it takes an ordinary camber form. In this way the form, on a few sections, is approximated. It is necessary, to prevent twist on the plates, that the sections be made parallel to A T.

Fairing-up the Form of the Protective Deck.—This may be done on the buttocks, but it is best to use diagonal lines like D E and F G. The cutting points of the trial form are lifted from the level line A H, and faired in the Sheer on the contracted method. After a few sections are faired, the intermediates can be easily run in by obtaining spots on the diagonals and heights for the centre from the Sheer, **Fig. 111**.

Expansion of the Protective Deck.—Assuming that the deck is a continuous fair surface, the general method is to girth the centre line $U^1 Y^1$ in **Fig. 113**—the Sheer enlarged for the fore end—for the position of the frame stations, stern, and stem terminations. Set these along a straight line representing the centre line of the vessel, and erect perpendiculars from the points, which are the expanded frame stations. On these lay off the girthed width, on corresponding sections, of the deck from the centre line: a curve through the spots will give J W X Y in **Fig. 114**. Repeat the process for the other side of the vessel. Upon this expansion all the openings, longitudinal

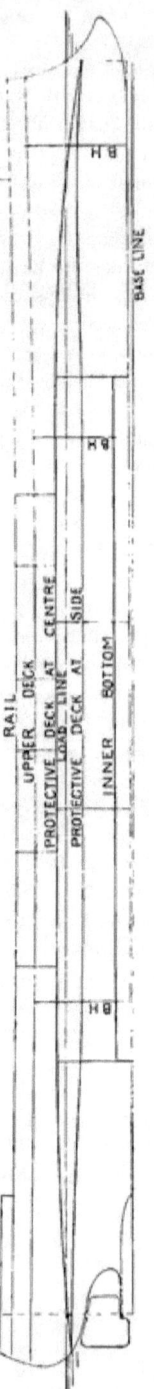

Fig. 111.
SHEER.

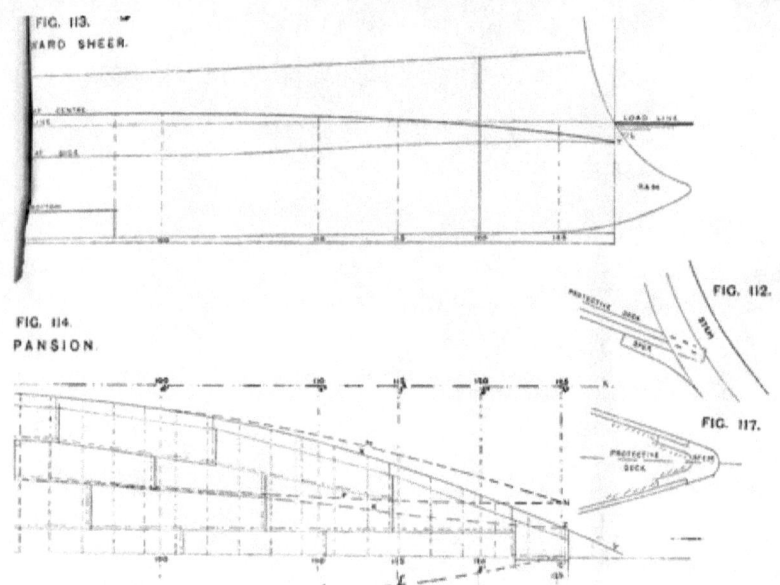

FIG. 113.
...ARD SHEER.

FIG. 114.
...PANSION.

FIG. 112.

FIG. 117.

ARMOUR.   117

and transverse bulkheads, are indicated before arranging the plate
edges; which are made parallel to the centre line over the horizontal
portion, and about parallel to the ship's side on the slope. The
edges must be straight between the butts. It will be evident that
some allowance in ordering the plates should be made, outside of that
for planing, owing to the expansion being made to the under side of
the thick plating. The beams, and whatever comes in contact with
the deck, are drawn in position. Those on the under side in blue ink,
and those on the top side in red.

A More Correct Method of Protective Deck Expansion.—
Draw in the Body, **Fig. 110**, a C and A B square to A T. Produce
lines of each section to A B, shown dotted. Lift from C on C a the
position of g, f, e, d, c, b, and a, and lay them off from the centre of
the expansion on the ordinary spacing of the frames, $a^1, b^1, c^1, d^1, e^1, f^1$,
g^1, and C^1. Then girth the midship form in the Body from U to T,
and T to A. This laid out parallel from the centre line of **Fig. 114**
will give J K. The correct position of the stations, on this line, will
be got by girthing L C^1, and laying them out along J K from J, and
from the points erect perpendiculars. Now girth each section, in the
Body, from A B to the side line A S, and the knuckle point T S, and
set off these girths below J K on their corresponding frames, which
will give J W M N, the expanded edge of the deck at side; and the
knuckle edge O P N. Next expand the deck centre, as before explained,
for the true position of the stations, and place the points forward of
67 on the expansion centre, and erect perpendiculars; on which plot
off the girthed widths of each section, from the centre to the knuckle
line T S, and you get line O R Z, from which it is seen that the surface,
when properly developed, parts at the ends in way of the knuckle. It
will be found that if M N P is traced neatly with the position of the
frames, it will fit into R Z, only the line of the frame stations will
appear curved.

This method is adopted, on the loft floor, to secure the true shape
of the deck at the side, so that templates may be made for the platers.
At the extreme ends, surface templates should be taken from the ship.

Fig. 114 shows only the fore end expanded; it is almost needless
to say that the after end is done in the same manner.

Method of Plating the Protective Deck.—This deck is plated
as described in **Figs. 114, 115,** and **116**. The thickest plating is on
the slopes, terminating at the knuckle. The top plate is occasionally
stopped short of the ship's side about 15 inches, to allow for attaching
the upper frame knees.

and transverse bulkheads, are indicated before arranging the plate edges; which are made parallel to the centre line over the horizontal portion, and about parallel to the ship's side on the slope. The edges must be straight between the butts. It will be evident that some allowance in ordering the plates should be made, outside of that for planing, owing to the expansion being made to the under side of the thick plating. The beams, and whatever comes in contact with the deck, are drawn in position. Those on the under side in blue ink, and those on the top side in red.

A More Correct Method of Protective Deck Expansion.—Draw in the Body, Fig. 110, a C and A B square to A T. Produce lines of each section to A B, shown dotted. Lift from C on C a the position of g, f, e, d, c, b, and a, and lay them off from the centre of the expansion on the ordinary spacing of the frames, a^1, b^1, c^1, d^1, e^1, f^1, g^1, and C^1. Then girth the midship form in the Body from U to T, and T to A. This laid out parallel from the centre line of Fig. 114 will give J K. The correct position of the stations, on this line, will be got by girthing L C^1, and laying them out along J K from J, and from the points erect perpendiculars. Now girth each section, in the Body, from A B to the side line A S, and the knuckle point T S, and set off these girths below J K on their corresponding frames, which will give J W M N, the expanded edge of the deck at side; and the knuckle edge O P N. Next expand the deck centre, as before explained, for the true position of the stations, and place the points forward of G7 on the expansion centre, and erect perpendiculars; on which plot off the girthed widths of each section, from the centre to the knuckle line T S, and you get line O R Z, from which it is seen that the surface, when properly developed, parts at the ends in way of the knuckle. It will be found that if M N P is traced neatly with the position of the frames, it will fit into R Z, only the line of the frame stations will appear curved.

This method is adopted, on the loft floor, to secure the true shape of the deck at the side, so that templates may be made for the platers. At the extreme ends, surface templates should be taken from the ship.

Fig. 114 shows only the fore end expanded; it is almost needless to say that the after end is done in the same manner.

Method of Plating the Protective Deck.—This deck is plated as described in **Figs. 114, 115,** and **116**. The thickest plating is on the slopes, terminating at the knuckle. The top plate is occasionally stopped short of the ship's side about 15 inches, to allow for attaching the upper frame knees.

The method of forming the edges is seen in **Fig. 116**, and of the butts in **Fig. 115**; only the bottom thickness is attached to the beams. The fore and aft edges are straight between the butts, as shown in the expansion, **Fig. 114**.

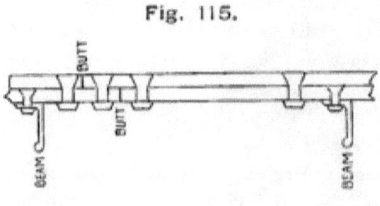

Fig. 115.

The **Figs. 112** and **117** explain the method of finishing on the stem. The after end is secured to the stern plating by an angle bar. This deck being continuous, all bulkheads are cut and connected to the deck by single bars. Owing to the numerous longitudinal bulkheads in the machinery space, which give support to the deck, only occasional strong beams are fitted there.

Model of the Protective Deck.—The most correct way to arrange and order the plating is to make a wood block model, on a ¼ inch scale, of the deck from or about the level plane A H in **Fig. 110**. This is rubbed smooth and painted white. The beams, bulkheads,

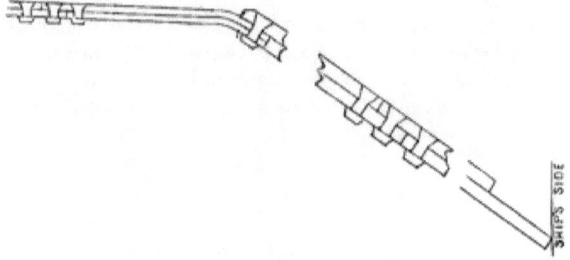

Fig. 116.

openings, etc., are marked upon it, and then the edges and butts of the plating are arranged. There should be no difficulty in ordering the plates from this model. At the irregular parts the correct surface of any plate may be got by tracing it, and then laying the tracing flat. Order the plate in the usual way described on page 50.

Beams Supporting the Protective Deck.—When this deck has much curve fore and aft, the beam flange is bevelled to suit.

ARMOUR. 119

The bevels are lifted *in the same manner* as already described for the frames of merchant vessels.

Belt Armour and Deck in a Battle Ship.—After the *moulded* form of the ship is faired-up on the loft floor, the position of the

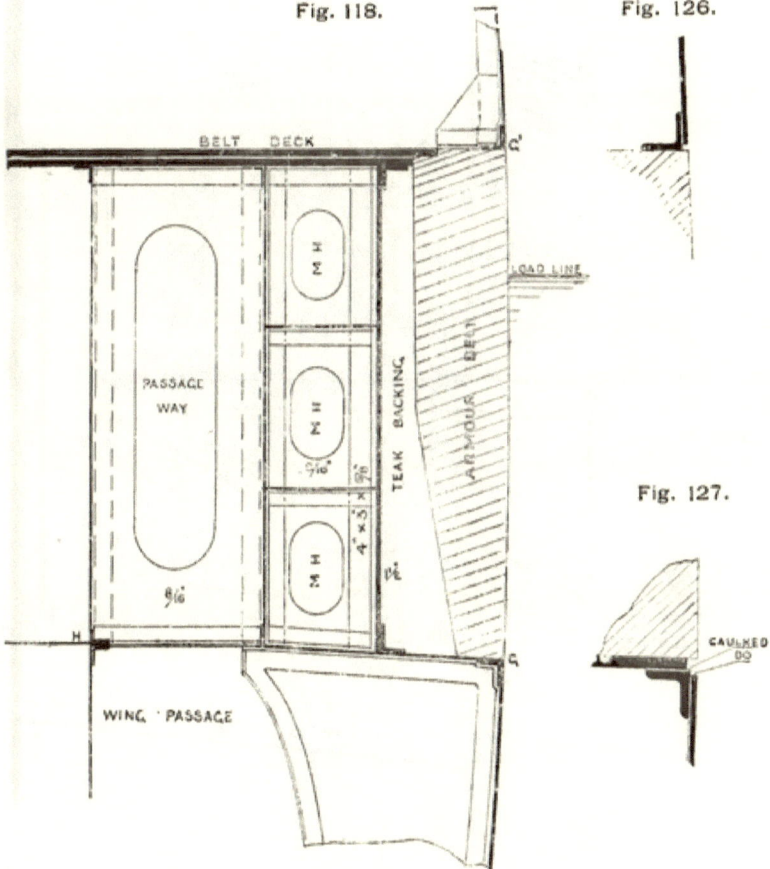

Fig. 118.   Fig. 126.   Fig. 127.

armour belt relative to the load-line is marked on the midship section, as shown in the enlarged sketch in **Fig. 118**, and also on the Body sections, **Fig. 119**. The armour will project out beyond the moulded

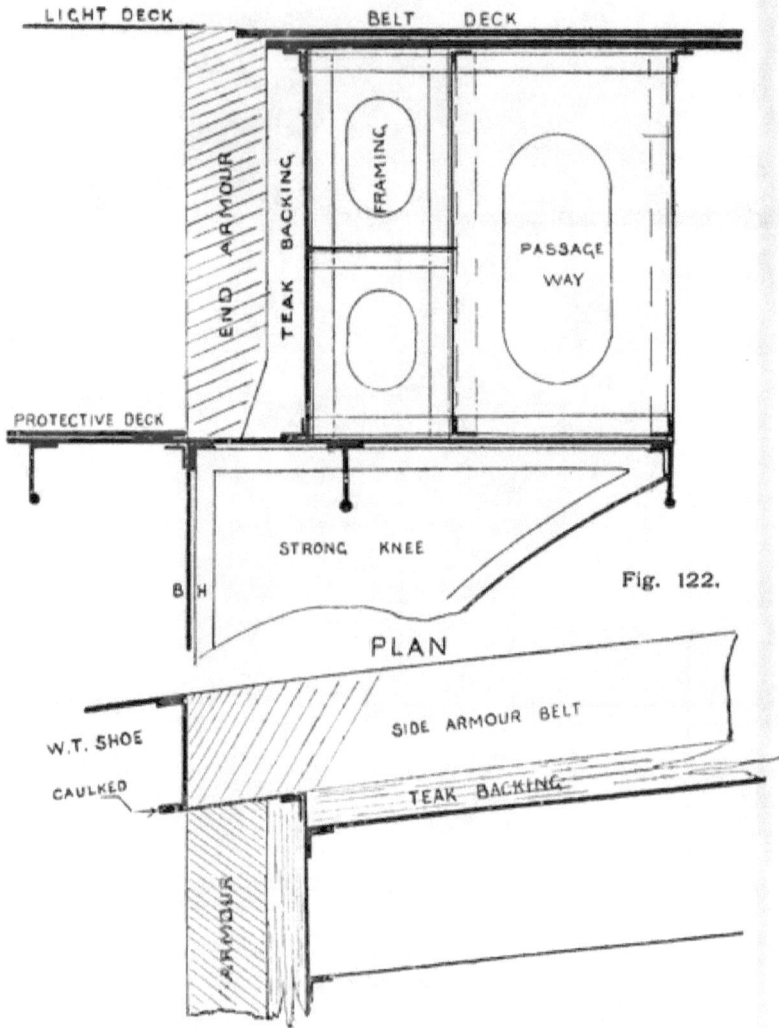

Fig. 120.

FORE AND AFT SECTIONAL ELEVATION OF END ARMOUR.

Fig. 122.

form the thickness of the adjacent shell plates, which are inside strakes, so that the extreme side will be flush for some distance. In fairing-up, on the loft floor, in the first case the moulded form, this projection is deducted; therefore, it must be added on in drawing in the armour. The armour belt extends at least over the length of the machinery space, magazines, and shell rooms. It is crowned by a level belt-deck, formed of two thicknesses of steel plating, reaching over the length of the armour, being connected at the sides and ends as shown in **Figs. 118** and **120**, and supported by strong bulb angle beams attached to alternate frames. The butts and seams of the plating are made the same as in **Figs. 115** and **116**. The side armour, which is not connected at the butts, is fixed against teak, or other hard wood, backing about 5 inches thick; which is supported by one or two thick continuous longitudinal steel plates riveted to the frame work as shown in **Fig. 118**. The entire system is called "Box framing." It is made level on the top and parallel to the ship's side. The frames next to the backing are double, as shown in the plan, **Fig. 123**. The deck or shelf, G H in **Fig. 118**, on which the armour rests, is inclined inboard to about 3 to 5 degrees from the side line G, which is *vertically* parallel to the top G^1. The lower edge of the armour is made to suit the bevel of this shelf. The section is the same thickness, in most cases, for about half its depth and tapered from that point, as shown in **Fig. 118**. At the extreme fore and aft ends it is slightly reduced in thickness. To begin the laying-off, the form of the armour with the frame work is drawn in at each section in the Body, **Fig. 119**, and faired-up on the floor by level or diagonal lines in the usual way. Where there is much fore and aft curve on the level lines, the thickness of the armour should be set in *square* to these outside lines in **Fig. 124**, and curves traced through the spots will give correct points, on the sections, for the inside edge.

When the armour belt only extends over the midship portion, as in **Fig. 129**, an armoured deck, called protective deck, is arranged forward and abaft of the belt, and run to suitable heights at the extreme ends; at the aft end to cover the steering gear, and at the fore end to give strength in ramming. This deck is a continuation of the armour shelf, but cambered up at the centre for greater efficiency. The armour, with the framework, is carried across the ship at the ends of the belt to form protection against fore and aft fire. It rests upon the protective deck extended, as shown in **Fig. 120** and cross section **Fig. 121**, supported by beams and a strong

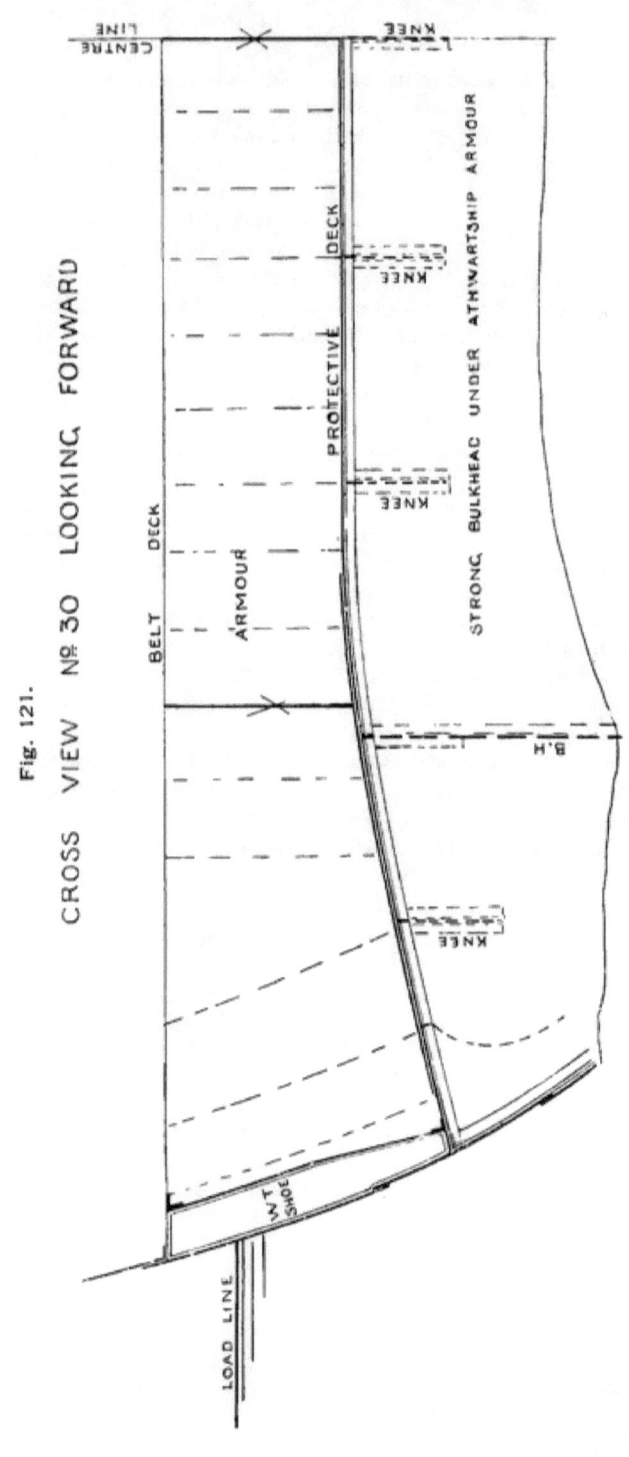

Fig. 121.

CROSS VIEW No 30 LOOKING FORWARD

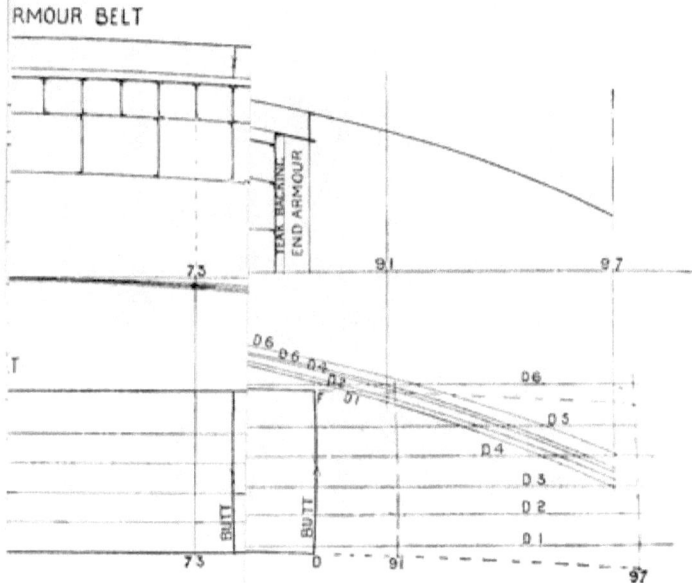

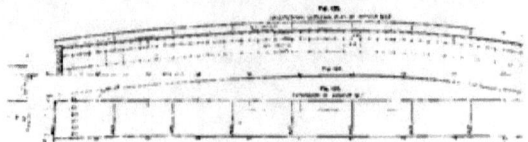

bulkhead with large brackets. The butts and edges of the armour are not caulked. The finish of the lower edge is shown by the **Fig. 127**, and of the top by **Fig. 126**.

The **Fig. 120** is a fore and aft sectional elevation through the end armour, and **Fig. 121** a cross section at 30 frame, showing how the belt and protective decks are finished and the armour supported. In **Fig. 122** is shown, in plan, the method of terminating the side and end armour with a watertight shoe. **Fig. 123** is a longitudinal sectional plan showing the disposition of the armour butts and box framing.

Fairing-up the Belt or Side Armour.—This may be done by close level lines, or straight diagonals, in the usual way. It is laid out in the **Figs. 119** and **124**, faired by six curved diagonals. No. 4 is first placed in the Body square to each frame curve, and the others are at an equal girthed distance each side of it. Then, taking the side line A B in Fig. 119 as a base, lift from it on the run of the diagonals the position of the frames, and set them off on their respective stations in the Fig. 124, and fair lines in. If the moulded form of the ship is fair, these ought to be correct. The inside of the armour and frame work may be faired in the same way, by setting in *square* to these lines the thickness of armour, and passing curves through the spots, which should be transferred on each frame and diagonal into the Body, and curves put through the points.

Expansion of the Belt Armour.—Girth on 67 frame in the Body, Fig. 119, the distance apart of the diagonals. Draw in **Fig. 125** a straight and parallel line to the base of the **Fig. 124**, and set off the position of the curved diagonals to it, which should be parallel distances apart. Then girth each diagonal in the **Fig. 124** for the position of the frame stations, and lay them off on their respective sides of the midship frame 67, of **Fig. 125**, and on their corresponding diagonals. A curve passed through these points will give shape of the frames in the expansion. The form of the lower edge may be found by girthing its position on each section, in the Body, **Fig. 119**, relative to No. 1 diagonal, and setting the distances off below D 1 on their frames in the expansion. A line drawn through the spots will give lower edge C D of the armour. Do the same with the top edge relative to D 6 and you get E F. It is necessary to find the form and position of the butts for ordering the plates, which you can do with relation to the frames, or they may be squared down on to the curved diagonals in the **Fig. 124**, and then girthed on each side and trans-

bulkhead with large brackets. The butts and edges of the armour are not caulked. The finish of the lower edge is shown by the **Fig. 127**, and of the top by **Fig. 126**.

The **Fig. 120** is a fore and aft sectional elevation through the end armour, and **Fig. 121** a cross section at 30 frame, showing how the belt and protective decks are finished and the armour supported. In **Fig. 122** is shown, in plan, the method of terminating the side and end armour with a watertight shoe. **Fig. 123** is a longitudinal sectional plan showing the disposition of the armour butts and box framing.

Fairing-up the Belt or Side Armour.—This may be done by close level lines, or straight diagonals, in the usual way. It is laid out in the **Figs. 119** and **124**, faired by six curved diagonals. No. 4 is first placed in the Body square to each frame curve, and the others are at an equal girthed distance each side of it. Then, taking the side line A B in Fig. 119 as a base, lift from it on the run of the diagonals the position of the frames, and set them off on their respective stations in the Fig. 124, and fair lines in. If the moulded form of the ship is fair, these ought to be correct. The inside of the armour and frame work may be faired in the same way, by setting in *square* to these lines the thickness of armour, and passing curves through the spots, which should be transferred on each frame and diagonal into the Body, and curves put through the points.

Expansion of the Belt Armour.—Girth on 67 frame in the Body, **Fig. 119**, the distance apart of the diagonals. Draw in **Fig. 125** a straight and parallel line to the base of the **Fig. 124**, and set off the position of the curved diagonals to it, which should be parallel distances apart. Then girth each diagonal in the **Fig. 124** for the position of the frame stations, and lay them off on their respective sides of the midship frame 67, of **Fig. 125**, and on their corresponding diagonals. A curve passed through these points will give shape of the frames in the expansion. The form of the lower edge may be found by girthing its position on each section, in the Body, **Fig. 119**, relative to No. 1 diagonal, and setting the distances off below D 1 on their frames in the expansion. A line drawn through the spots will give lower edge C D of the armour. Do the same with the top edge relative to D 6 and you get E F. It is necessary to find the form and position of the butts for ordering the plates, which you can do with relation to the frames, or they may be squared down on to the curved diagonals in the **Fig. 124**, and then girthed on each side and transferred into the expansion. The form of the end armour may be got by drawing out cross sections, like **Fig. 121**.

Moulds for Ordering the Belt Armour.—A surface mould is made from the expansion of each plate, showing the position of the frames and bolts for attaching the armour, together with moulds for the top, bottom, ends, and two vertical winding sight moulds for fixing on to the outside edge of the plate near each butt. Altogether seven moulds are required for each plate. The butts are vertical and square to the outside surface on the top edge.

Barbette.—**Fig. 128** shows the sectional plan and elevation of a forward barbette, standing on the belt deck, from which a big twin gun is worked. Its position is further seen in **Fig. 129**. The armour is supported by teak backing against one or two thick steel plates, riveted to the frame work—these plates take the armour bolts. The inner edge A B C is parallel to the outside of the armour D E F. The crown G H is level with the belt deck, and made of thick steel plates, which at the sides lie at an inclination of about 5 degrees to suit gun depression. An opening is arranged in the crown to allow for the training of the gun. The armour butts are made square to the outside surface to come on to the frames. The butts of the inner thick plates also come on frames. The seam edges are formed by lapping each thickness, as in **Fig. 130A**. Intercostal plates a, b, c, d, e, and f are fitted in between the frames, clear of the bolts, for support. The armour is slightly reduced in thickness at the after end from P P¹, and a joint K¹ L¹ arranged clear of the intercostals. Below K¹ L¹ the armour is considerably reduced in thickness. The forward end of the barbette is circular with the roller path, and the after end is circular with a suitable radius, while in between is a straight line. The inside surface of the thin armour may be made flush with the top thickness, which gives parallel backing.

Expansion of the Barbette Armour.—To draw an approximate expansion, show in the line R S T in the plan at half the thickness of the armour, and produce the direction of the frames to cut this line, then girth from R to T, marking on the lath, or paper, the position of the frames, butts, and centre line. Lay this girth R T, with frame positions, on a base R¹ T¹, **Fig. 130**, and erect frame lines. Set off G¹ H¹ parallel to R¹ T¹, and at a distance equal to the depth of the armour; also K L equal in position to K¹ L¹. Arrange the butts of the armour, and the thick plates on the frames, to clear each other; and the edges of the thick plates to clear the intercostals and armour

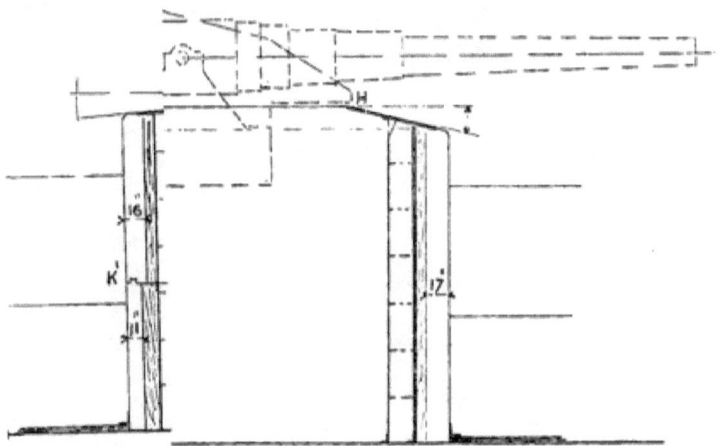

FORE & AFT SECTIONAL ELEVATION

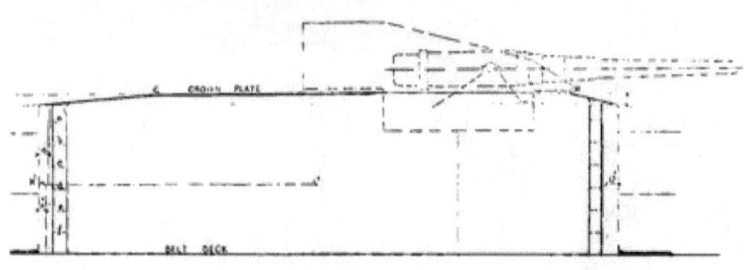

PLAN

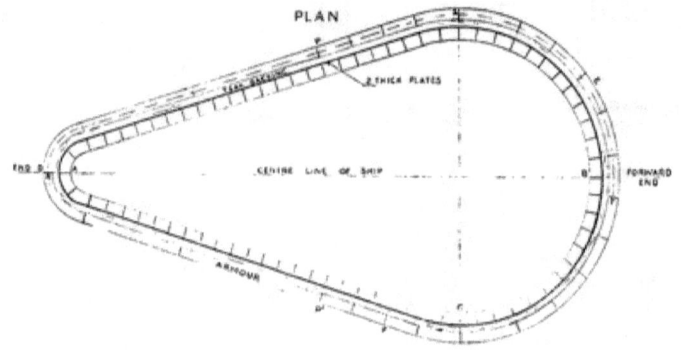

Fig. 135
BATTLE SHIP
SHEER

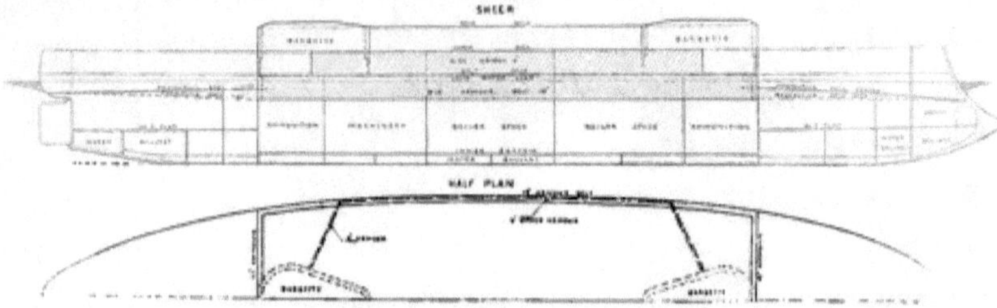

HALF PLAN

Fig. 130.

EXPANSION OF ARMOUR FOR ONE SIDE.

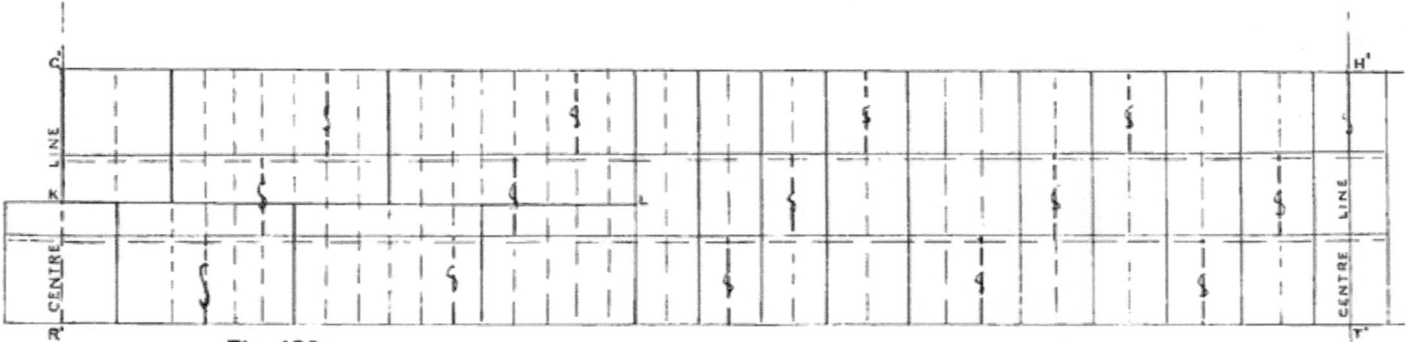

Fig. 130A.

Fig. 130.

EXPANSION OF ARMOUR FOR ONE SIDE.

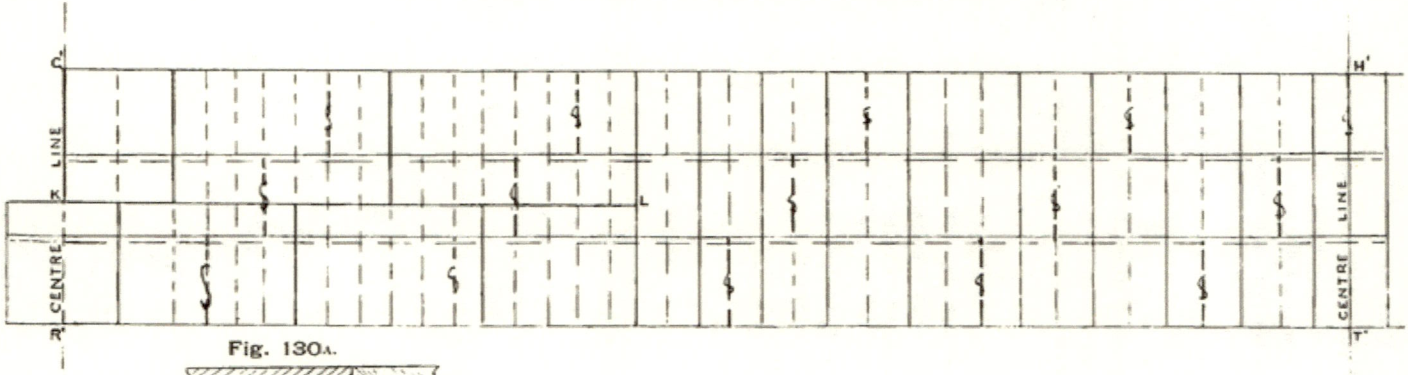

Fig. 130A.

bolts. This plan, which shows one side of the barbette, is suitable for yard purposes, or may be used for obtaining an approximate account of the weight of the armour.

Ordering the Barbette Armour.—The barbette is mocked up, built in skeleton fashion, on the loft floor and wood moulds made therefrom. One for each end of every plate, and flat moulds for the bottom and top, and one for the outside surface, with two vertical sight winding pieces for the outside to get the *round* plates into form. The inner plating may also be measured from the loft at the same time, due allowance being made for planing edges and butts, or a correct account of the armour for ordering may be got by expanding on the left floor the outside surface D E F, and supplying the moulds before-mentioned. A correct account of the inner thick plates may be got by expanding the surface between the plates in the usual manner.

Circular Barbette.—They are sometimes arranged in complete circular form, as in **Fig. 131**, with a supply armoured tube from magazines, etc., placed in the centre of the roller path B B. An account of this barbette armour is easily found by laying the plan form down on the loft floor, and producing the direction of the frame stations to the outside surface line. Girth this line for the stations, and lay them out on a base, and erect station perpendiculars. Show in the depth, which will be parallel to the base. This may be got by producing the top and side lines in the section until they meet. This surface expansion is sufficient, with the ends, top, bottom, surface, and sight moulds of each plate, for ordering the armour. The butts are on the frames as before explained.

Revolving Armoured Turret or Redoubt.—An armoured revolving turret carrying guns is occasionally fitted inside of a barbette of the form of **Fig. 130** or **Fig. 131**. In **Fig. 132** is given an outline sketch showing such, which needs no further explanation. The armour, framing, etc., being arranged and expanded in the same way as the barbette previously mentioned.

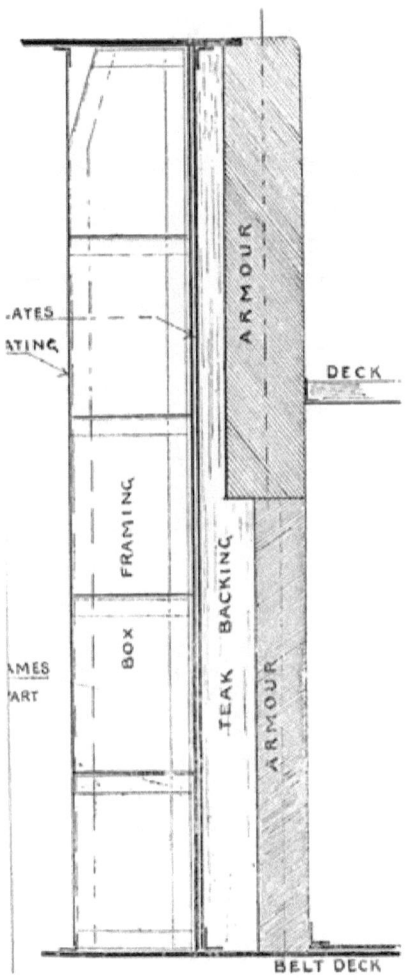

ENLARGED SIDE SECTION

Fig. 131.

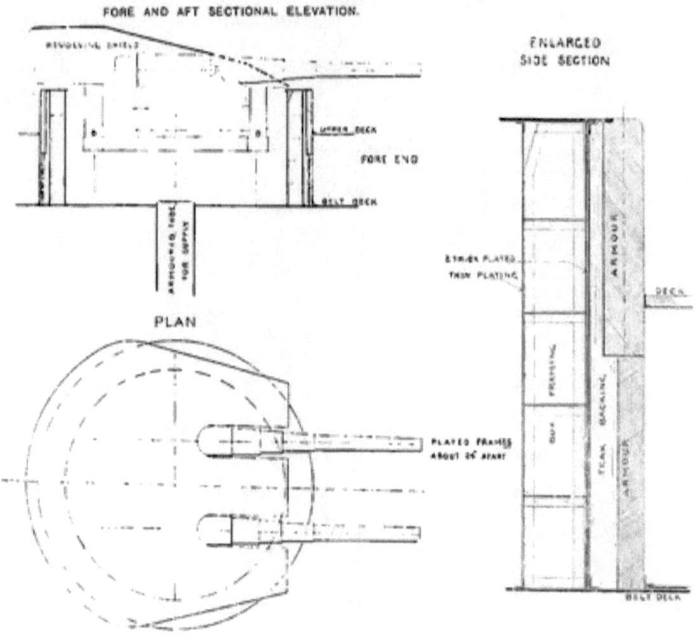
FORE AND AFT SECTIONAL ELEVATION.
ENLARGED SIDE SECTION
UPPER DECK
FORE END
BELT DECK
PLAN
ARMOUR
DECK
TEAK BACKING
ARMOUR
BELT DECK

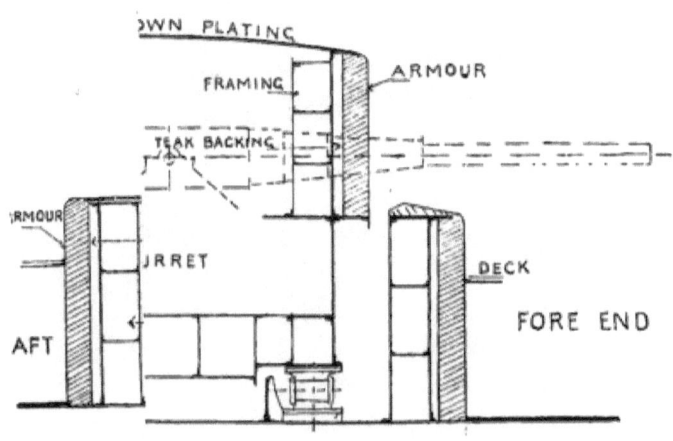
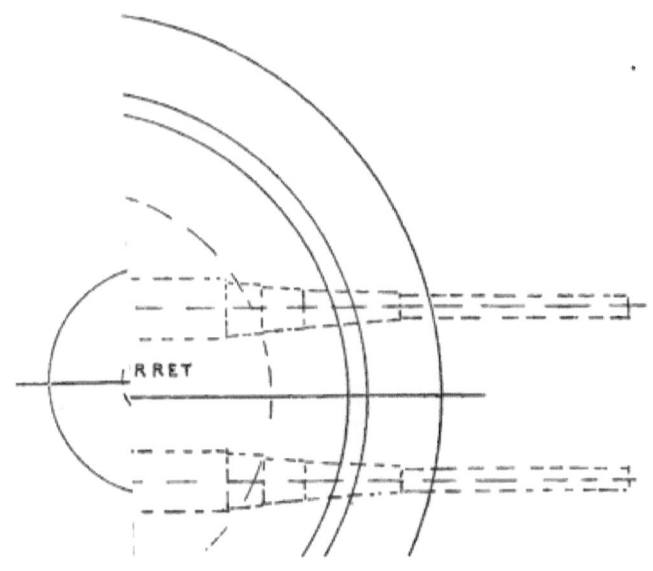

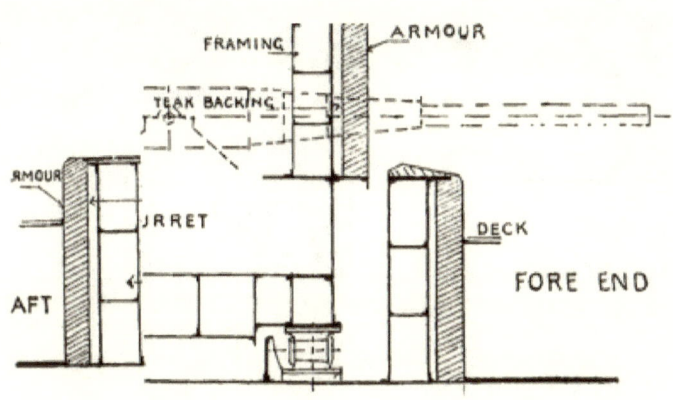
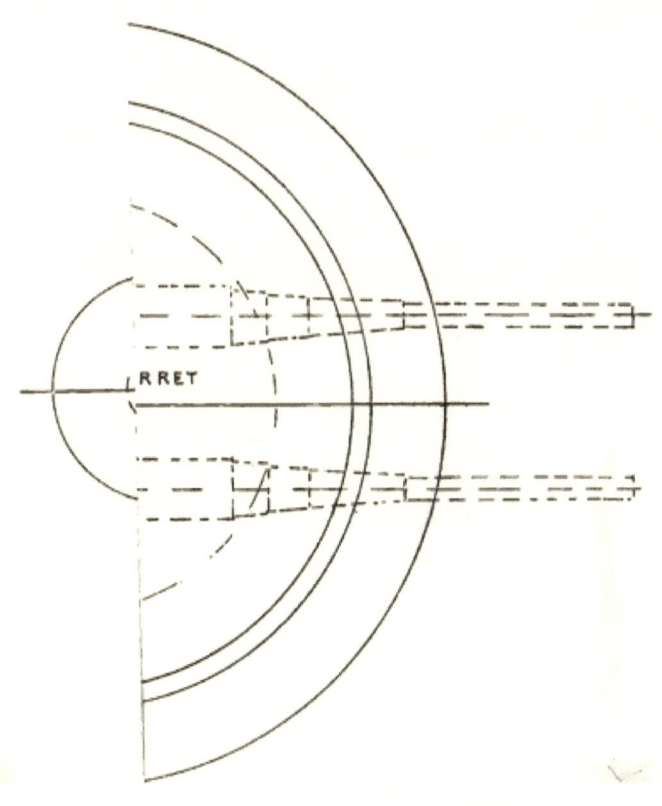

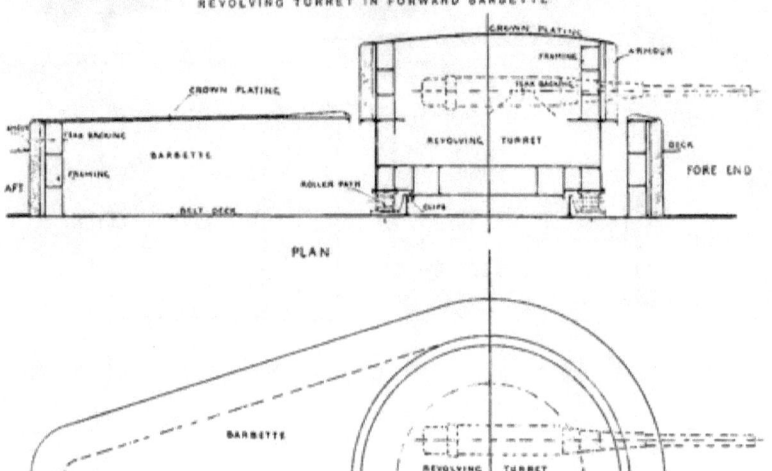

Fig. 152.
FORE & AFT SECTIONAL ELEVATION.
REVOLVING TURRET IN FORWARD BARBETTE.

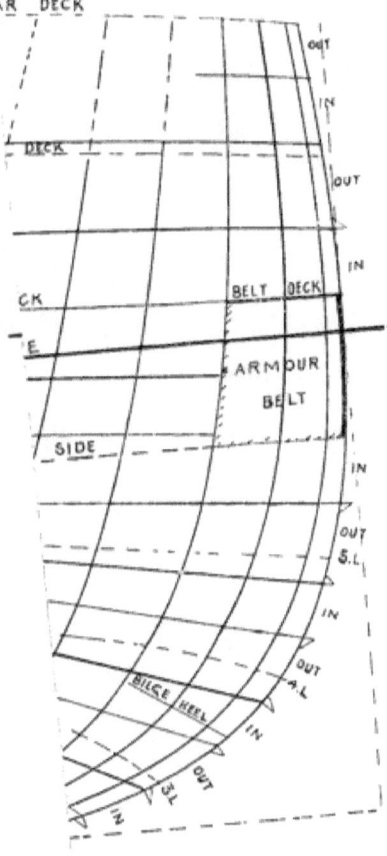

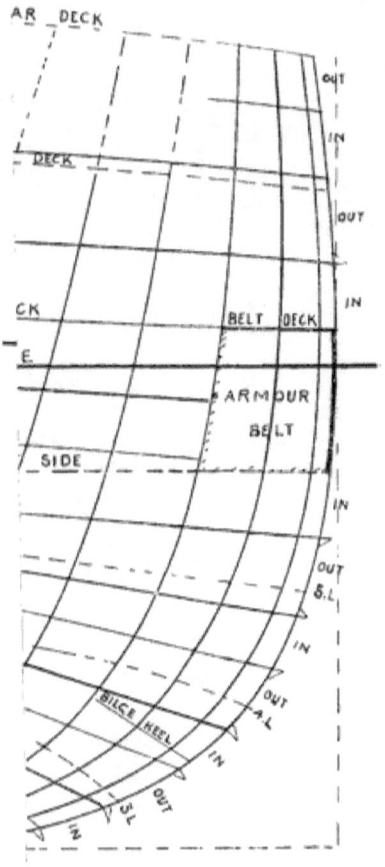

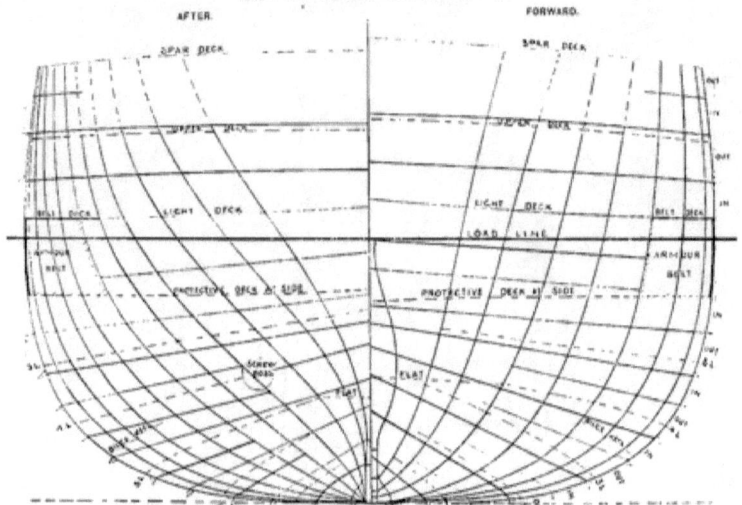

Fig. 133.
BODY SECTIONS OF A BATTLE SHIP.

CHAPTER XIV.

Outer Bottom Plate Edges of a Battle Ship—Outer Bottom Plate Edges of a Cruiser—Bilge Keels: where best Placed—Bossed Frames Forward in way of the Ram.

SHELL PLATING AND BILGE KEELS.

Outer Bottom Plate Edges of a Battle Ship.—These are first arranged, approximately, on the Body sections as in **Fig. 133**, in conjunction with the inner bottom longitudinal sight edges: and afterwards lined off and faired on the model.

In **Fig. 133**, show in the extent and position of the side armour and deck lines. Set off around the midship section of each Body the sight edges of the shell plates, which are made of a suitable width somewhere about 50 inches, or they are lifted from the "scantling section." The flat plate keel is an outside strake, divided equally each side of the centre line, with a doubling or double plate keel inside. The strakes next to the top and bottom of the armour belt should be inside and flush with it. Next draw in the position of the screw boss, the flats at the extreme ends, and, it is best to approximate, the lines for the inner bottom longitudinals as explained on page 130, so that the shell edges may be kept clear—they are indicated by dotted lines. The edges of the shell plating, above the armour belt, are sheered to the upper deck line, with a slight taper towards the ends. Those below the belt are straight to the stem and stern, with considerable taper, and made to clear inner bottom longitudinals "Stealers" are introduced to prevent narrow strakes. Under the extreme bottom a slight round is given to form a better line. The plate keel is of parallel girth for the best part of the vessel's length, and lifted up at the ends to take away the appearance of drooping. The parts before and abaft of the belt are divided in suitably; the upper edge of the strake below the armour may be stepped up to make a more pleasing line and division, and to get it clear of the protective deck side. Care should be taken at the screw boss to arrange the edges for one furnaced plate, or to make a plate edge on the centre of the round, and have two furnace plates as shown in **Fig. 134**.

Edges should clear all watertight flats. After arrangement in the Body, they are lifted and faired on the model in the manner described

CHAPTER XIV.

Outer Bottom Plate Edges of a Battle Ship—Outer Bottom Plate Edges of a Cruiser Bilge Keels; where best Placed—Bossed Frames Forward in way of the Ram.

SHELL PLATING AND BILGE KEELS.

Outer Bottom Plate Edges of a Battle Ship.—These are first arranged, approximately, on the Body sections as in **Fig. 133**, in conjunction with the inner bottom longitudinal sight edges; and afterwards lined off and faired on the model.

In **Fig. 133**, show in the extent and position of the side armour and deck lines. Set off around the midship section of each Body the sight edges of the shell plates, which are made of a suitable width somewhere about 50 inches, or they are lifted from the "scantling section." The flat plate keel is an outside strake, divided equally each side of the centre line, with a doubling or double plate keel inside. The strakes next to the top and bottom of the armour belt should be inside and flush with it. Next draw in the position of the screw boss, the flats at the extreme ends, and, it is best to approximate, the lines for the inner bottom longitudinals as explained on page 130, so that the shell edges may be kept clear—they are indicated by dotted lines. The edges of the shell plating, above the armour belt, are sheered to the upper deck line, with a slight taper towards the ends. Those below the belt are straight to the stem and stern, with considerable taper, and made to clear inner bottom longitudinals. "Stealers" are introduced to prevent narrow strakes. Under the extreme bottom a slight round is given to form a better line. The plate keel is of parallel girth for the best part of the vessel's length, and lifted up at the ends to take away the appearance of drooping. The parts before and abaft of the belt are divided in suitably; the upper edge of the strake below the armour may be stepped up to make a more pleasing line and division, and to get it clear of the protective deck side. Care should be taken at the screw boss to arrange the edges for one furnaced plate, or to make a plate edge on the centre of the round, and have two furnace plates as shown in **Fig. 134**.

Edges should clear all watertight flats. After arrangement in the Body, they are lifted and faired on the model in the manner described on page 45, any correction necessary being transferred back into the Body. They are further faired-up on the loft floor before scrieving down (see page 48). The butts and thicknesses of the plating are set off on a ¼ inch scale "expansion plan" explained on page 48, and made to clear the butts of the armour, deck stringers, and longitudinals. They are afterwards transferred on to the model and adjusted if necessary.

Outer Bottom Plate Edges of a Cruiser.—These are arranged on the Body (**Fig. 134**) in a similar manner to a merchant ship, set forth on page 45. The flat keel plate is an outside strake and of parallel girth for almost the full extent, only at the extreme ends it is lifted up to form a better line. Before arranging plate edges, the position of the decks, longitudinals, and screw-boss should be indicated on the Body. The strakes above the protective deck are

Fig. 135.

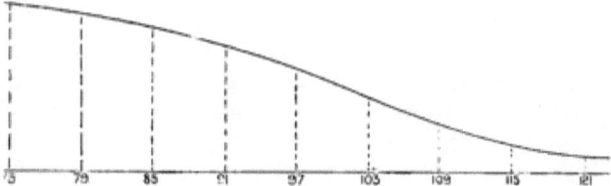

sheered to the upper deck line, and tapered slightly towards the ends; the taper being greatest upon those strakes nearest to the protective deck. The upper deck sheer strake stands about 9 inches above the deck line, slightly increasing towards the ends. The lower strakes, except the keel plate and garboards, are drawn in straight from the midship section to the stem and stern, about square to the frame curves, to divide the space to the best advantage. Owing to the fineness of the ship's form, it is necessary to introduce several "stealers" or "lost strakes" to prevent narrow plating. When all the lines are drawn fair in the Body, their position is girthed on each section on narrow strips of paper and transferred on to the model, and faired in the usual way. They are also tried on the loft floor before scrieving in.

Bilge Keels are placed on the turn of each bilge, being the points of least resistance to the vessel's speed and that of greatest

RUISER.

FORWARD.

Fig. 134.
BODY SECTIONS OF AN UNARMOURED CRUISER.

FORWARD

efficiency against rolling. The line is shown in each Body, in **Figs. 133** and **134**, drawn as square as possible to frame curves throughout its length: which is about half length amidships. This line appears curved when placed in the Sheer and Half Breadth.

Bossed Frames Forward.—It may be as well to note here, that sometimes the fore end frames are bossed for some distance from the stem, in a line with the centre of the ram. Particular attention should be given to the line through the centre of the boss, to make it graceful and easy. Such a line is shown in **Fig. 135**. The manner of obtaining and fairing is on the method explained on page 12, when the initial centre line is settled.

efficiency against rolling. The line is shown in each Body, in **Figs. 133** and **134**, drawn as square as possible to frame curves throughout its length; which is about half length amidships. This line appears curved when placed in the Sheer and Half Breadth.

Bossed Frames Forward.—It may be as well to note here, that sometimes the fore end frames are bossed for some distance from the stem, in a line with the centre of the ram. Particular attention should be given to the line through the centre of the boss, to make it graceful and easy. Such a line is shown in **Fig. 135.** The manner of obtaining and fairing is on the method explained on page 12, when the initial centre line is settled.

CHAPTER XV.

To Obtain and Fair the Lines of the Inner Bottom of a Cruiser—Expansion of the Inner Bottom—Expansion of the Longitudinals—Expansion of a Longitudinal on Curved Diagonals—Mocking-up System of Expansion—To Obtain and Fair-up the Inner Bottom Lines of a Battle Ship—Expansion of the Inner Bottom—Bevels on Inner Bottom Frames.

DOUBLE BOTTOMS.

It is, almost, an invariable rule to fit war vessels with a double bottom on the cellular system for at least the length of the machinery space, with ordinary framing at the ends.

To Obtain and Fair the Lines of the Inner Bottom of a Cruiser.—After, or at the same time as, the shell plating sight edges are settled, the curve of the inner bottom, A B C and A B'C', on the midship section, is transferred from the "scantling section" on to the large sized Body sections (**Fig. 136**), usually $\frac{1}{8}$ an inch = to 1 foot. These sections are got out, at convenient distances apart, for the full extent of the double bottom. Sufficient allowance is made on B C and B'C' to get the frame bar through, and the watertight collar fitted comfortably round it. The shell edge, $C^1 C^3$ and C and C^2, is run in, either in the Sheer or on the model, clear of the outer bottom edges, if possible, and the line faired-up and corrected in the Body. The intersection of this line with each frame will be the point of the inner bottom termination at the sides. Draw level lines through each point like B C and B'C', about the midship width, and fair the inner edge in the Half Breadth. The heights of the inner bottom fore and aft on the centre line are usually fixed on the "keelson plan," and lifted therefrom at each section, and set above the base on the centre line of the Body. The sight edges of the longitudinals, E D F and O M N, should now be drawn in clear of the plate edges, as far as possible, and placed about the middle of a shell strake. They are faired in the Half Breadth by lifting the cutting points from the Body square to the centre, and transferring (see F D E in **Fig. 137**). Then, from the cutting points of F D E, in **Fig. 136**, draw in the longitudinal square to each frame and about the midship width, or of a graduated specified taper ; and through these points, in conjunction with the centre heights and the knuckle, approximate curves of the inner bottom may be

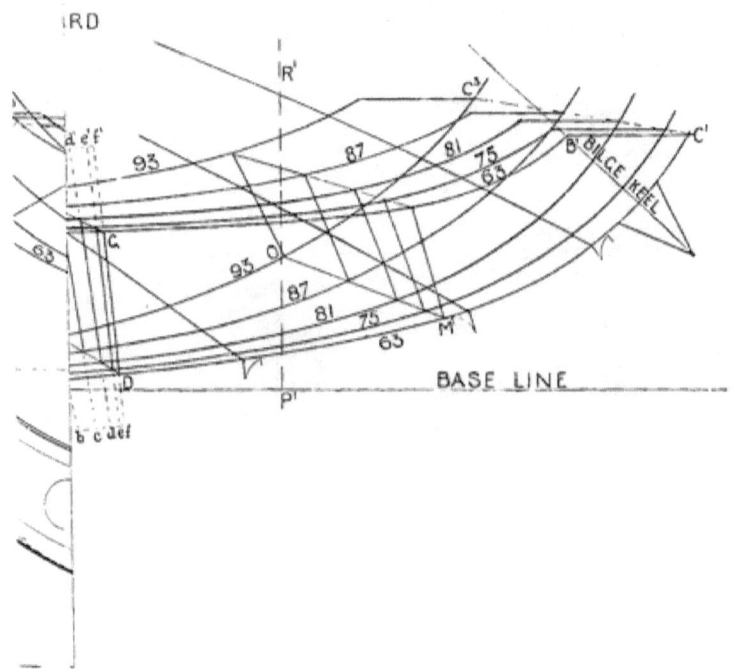

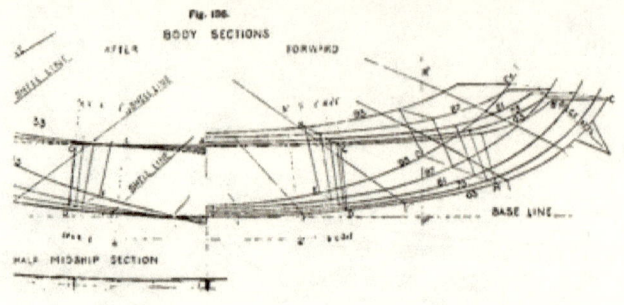

Fig. 106.
BODY SECTIONS

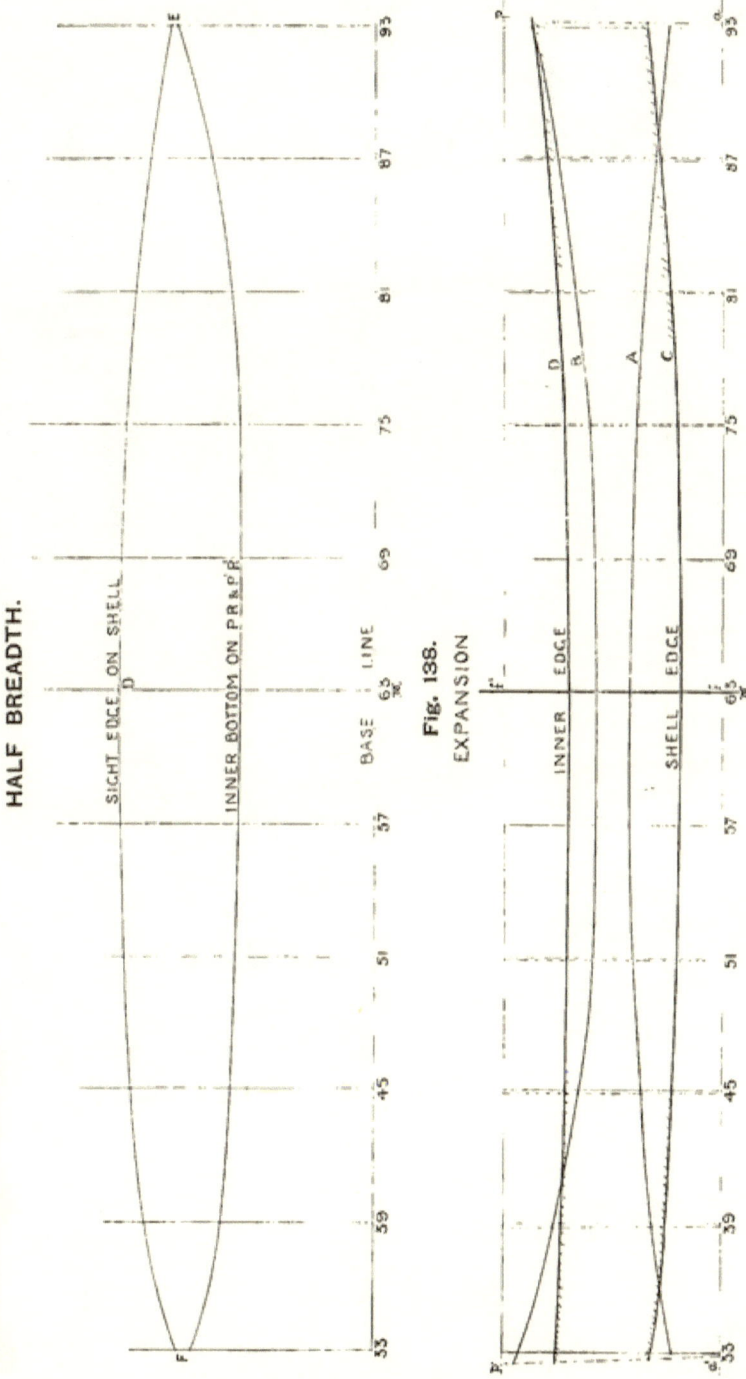

Fig. 137.
HALF BREADTH.

Fig. 138.
EXPANSION

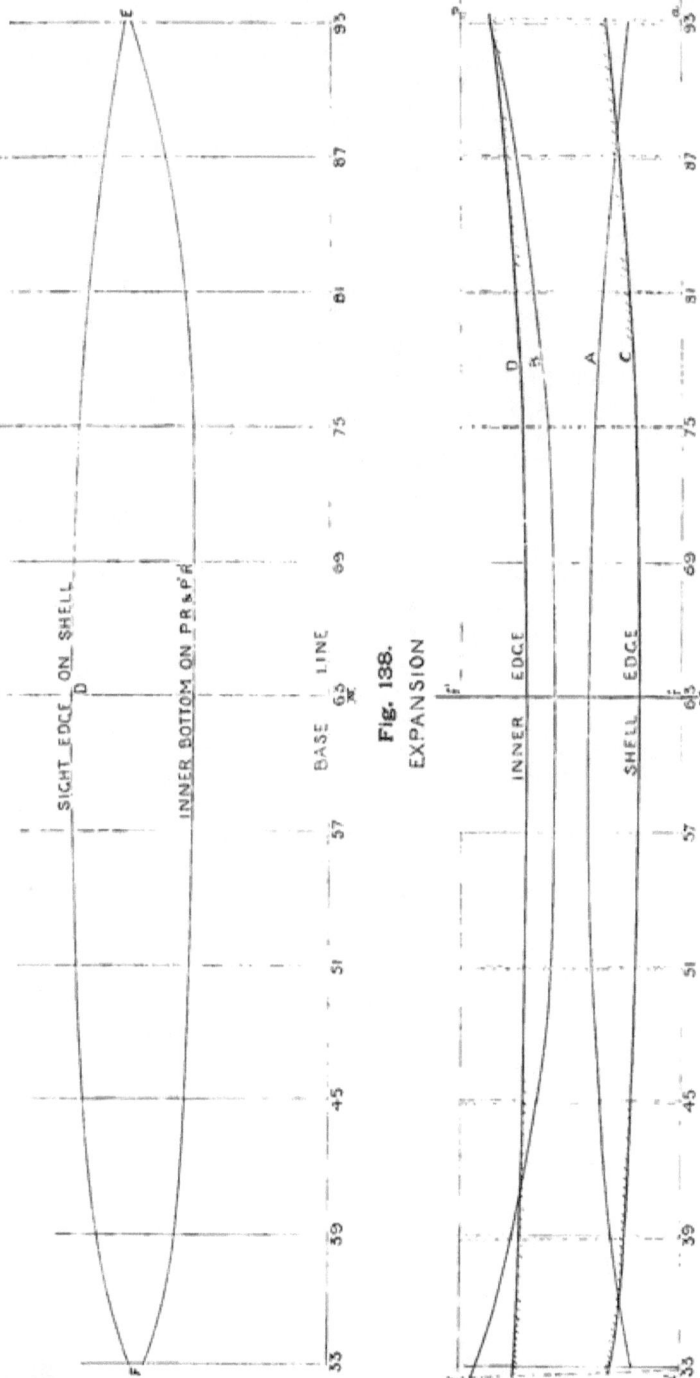

Fig. 137.
HALF BREADTH.

Fig. 138.
EXPANSION

drawn. Other longitudinal girders may now be shown in the same manner. You are now in a position to test the fairness of the inner bottom, which will, undoubtedly, require some modification. Sections 33, 63, and 93 should be considered fixed points, and the intermediates faired, which may be done on the buttocks by lifting the points, x, t, w, x, y, etc., on P R and P¹ R¹, from the base and transferring the heights into **Fig. 137**, or diagonals square to the curves may be used. It is usual to do this on the contracted method.

The top edge of the longitudinal, H G L, is faired by lifting the distances square out from the centre, and transferring them into the Half Breadth. They should require little or no alteration, seeing the longitudinals are square to the frames. It will be evident that the intermediate inner bottom points for the serieve board are lifted from these faired lines. Another method, differing slightly, is explained on page 135.

Expansion of the Inner Bottom in a Cruiser.—After the fairing of the sections is completed, bend laths, or strips of drawing-paper, round the lines of the inner bottom curves in **Fig. 136**, and mark on them the position of the centre, longitudinals, knuckle, and the shell point. On a base, representing the length of the double bottom on the same scale, erect the frame stations, and set off on their corresponding frames the girthed distances, and fair curves through the spots, which will give you the expanded form. Then girth the position of the edges of the inner bottom plating from the "scantling section," or fix them suitably to clear longitudinals, and lay them off from the expansion centre to the same scale ; producing lines through the points, parallel to the centre. Care should be taken to avoid at the ends feather-edged plates, and also to keep the edges clear of the watertight longitudinals. It is best where overlapping cannot be avoided to adopt the step system, or introduce "stealers." The position of the shell butts, manholes, engine and boiler seating, longitudinal and transverse bulkheads, watertight divisions, tunnels, openings, and whatever has to come in contact with the plating, is distinctly marked in coloured ink : then, the butts are arranged clear of those of the shell and longitudinals ; and the plates ordered. Owing to the round fore and aft, a little should be allowed on the length of the plates, or the correct position of the frames can be found in the first case by expanding longitudinally.

Expansion of the Longitudinals in a Cruiser.—In **Fig. 138**, draw down straight line $a^1 a$, representing the length of the tank on the same scale as the Body sections ; and mark off the position of the sections, from which points erect perpendiculars. In **Fig. 136** make $a\,p$ and $a^1\,p^1$ of equal length, and at equal distances, and parallel to D G and D G respectively on each side, $a\,a^1$ being parallel to the base line. From points a^1, p^1, a, and p, erect perpendiculars, and produce the line of the longitudinals at each frame to meet these in the points l, k, h, g, f, and so on, shown by dotted lines. In expansion (**Fig. 138**) set off $a^1\,f\,a$ and $p^1\,f^1\,p$ parallel, and distance $a\,p$ apart, taken from **Fig. 136**. Then lift from a and a^1 the points $b, c, d, e, f, g, h, i, k, l$, and set them on their respective frames above $a\,a^1$, in **Fig. 138**; through which draw curve A. Lift from p and p^1 the points b^1, c^1, d^1, etc., and lay them off on their respective frames below $p^1\,p$, in **Fig. 138**, and draw B. Girth A with a batten for the position of the frames, and place the batten on $a^1\,a$, keeping G3 fair with G3, and mark expanded position of the frames. Girth B likewise, and lay off the points around G3 frame on $p^1\,p$. Join the top and bottom spots, shown by dotted lines. From $a\,f\,a^1$, in **Fig. 136**, lift the position of the line F D E and L G H, on the run of the longitudinal, and mark them off on the expanded frame stations, in **Fig. 138**, from $a^1\,a$, which will give lines C and D. This is approximately the expanded form of the longitudinal. If the lines D G, E H, and the intermediates, were parallel to each other, this would be a correct development. The method just described is found sufficiently correct for the midship body, where the lines of the longitudinal at each section are about parallel to one another. When they are not quite parallel, which is an evidence of twist, if the expansion is taken in short lengths to embrace only one plate, there is not much error involved. At the extreme ends the lines are usually considerably out of the parallel, indicating great twist on the plate ; in such a case the mocking-up system explained further on is adopted.

Expansion of a Longitudinal by Curved Diagonals.—In **Fig. 139**, the Body, A G is the sight edge and $a\,g$ the inner edge of the longitudinal. Produce lines A a, B b, C c, etc., indefinitely in each direction. Fix a point a^1 for curved diagonal $a^1\,g^1$, and make it square to each section in the points a^1, b^1, c^1, etc. From A and b draw A L and $b\,k$ parallel to $a^1\,g^1$, by setting off $a^1\,b$ to the right of the points a^1, b^1, c^1, and a^1 A to the left, on the run of the longitudinal sections. Number these diagonals 1, 2, and 3. Girth each from the base A b for the points a^1, b^1, c^1, etc., and lay them off on their corresponding stations in the **Fig. 140**, through which draw fair lines.

Make $L^1 A^1$, in **Fig. 141**, parallel to H K. Set off $g^2 a^2$ and $h^1 k^1$ equal in distance to A a^1 and $a^1 h$ respectively in the Body, and parallel to $L^1 A^1$. Girth each diagonal from 67 frame in the **Fig. 140** for the frame stations, and lay them out on their corresponding lines from No. 67 in the expansion **Fig. 141**—No. 1 diagonal on $A^1 L^1$, No. 2 on $a^2 g^2$, No. 3 on $h^1 k^1$. Draw vertical curves through the spots, which will give form of the frames. Upon these indicate the position of the sight edge A G, relative to $a^1 g^1$ in the Body, i.e., lift a^1 A, b^1 B, c^1 C, d^1 D, e^1 E, f^1 F, g^1 G. Set off a^1 A, b^1 B, c^1 C, d^1 D on their respective frames, 67, 55, 49, 43, below $a^2 g^2$, and e^1 E, f^1 F, g^1 G,

Fig. 142.

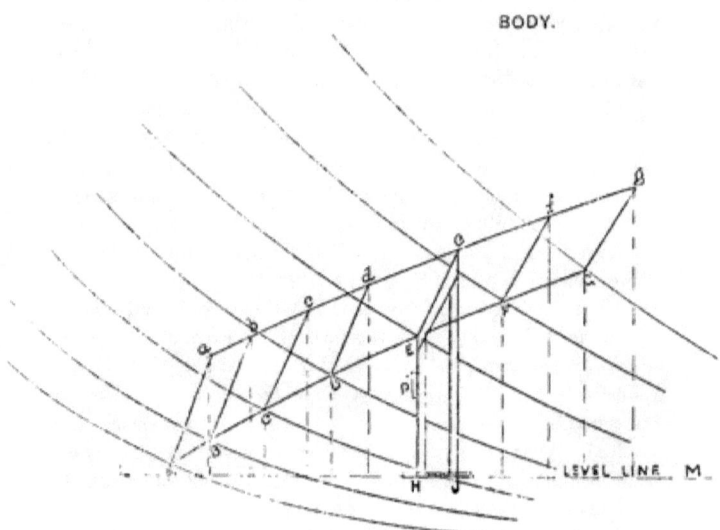

above $a^2 g^2$ on corresponding frames 37, 31, 25. A curve through the spots will give longitudinal sight edge $A^1 G^1$. In the same manner lay off $a g$ relative to $a^1 g^1$. The points a, b, c will be below, and d, e, f, g above $a^2 g^2$. Then the lines $a^2 g^2$ and $A^1 G^1$, **Fig. 141**, show the expansion of the surface of the longitudinal A G $g a$.

All the frames and butts are now shown across, and the plates ordered. The method comes very near to accuracy, if carefully carried out. Only the after end is shown, but it will be evident that the fore end can be done in the same way.

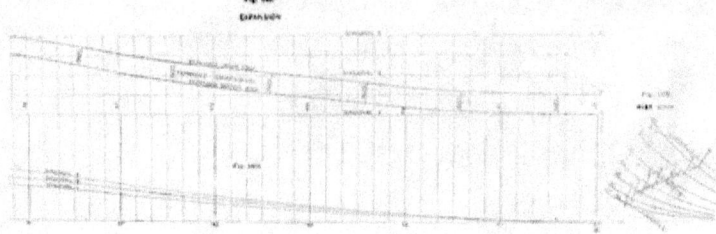

Expansion by the Mocking-up System.—Fig. 142 shows part of the Body plan on the loft floor with longitudinal sight edge and the intersection form at various stations. Drop perpendiculars from the points A, B, C, D, etc., and from a, b, c, d, etc., on to a convenient level line L M. Transfer the sight edge A G into the Half Breadth, lifting the cutting points square out from the centre line. Then make skeleton templates at each section showing the form, like H E e J. Erect these in the Half Breadth on their respective stations with the edge of the mould, A, B, C, D, E, F, and G, fair to the sight edge just transferred from the Body. Secure them, perpendicular, by a light cross piece tacked at the bottom to the floor, and attach them to each other by a fore and aft piece, P. The top outside edges of these templates give the *correct* form of the longitudinal surface and the position of the frames. A mould, like **Fig. 143**, is then made for each plate showing the extreme edges,

Fig. 143.

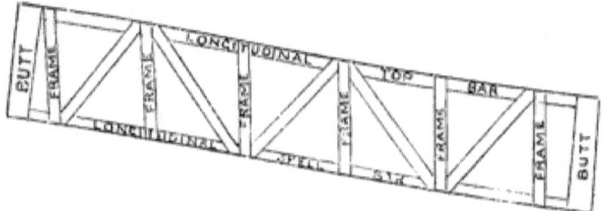

with cross pieces for the position of the frames, and end pieces for the butt straps. The latter should be arranged for the strap to clear the vertical bars. This is the method generally adopted where there is considerable twist, and absolute correctness is required. The watertight longitudinal butts are usually treble riveted and the non-watertight double.

NOTE.—It may be noted that an easy way of transferring a line from the loft floor on to a board for making a template is to place small flat-headed nails with the edge of the head resting on the line, and then lay a board on the top, and a little pressure will cause them to adhere to the board, when it may be lifted and the curve drawn in pencil through the points.

134

Make
in di
L¹ A¹
frame
No. 6
a² g²,
will
the s
c¹ C,
respe

above
spots
lay of
f, g a
expan
  A
ordere
carrie
that

Expansion by the Mocking-up System.—Fig. 142 shows part of the Body plan on the loft floor with longitudinal sight edge and the intersection form at various stations. Drop perpendiculars from the points A, B, C, D, etc., and from *a*, *b*, *c*, *d*, etc., on to a convenient level line L M. Transfer the sight edge A G into the Half Breadth, lifting the cutting points square out from the centre line. Then make skeleton templates at each section showing the form, like H E *e* J. Erect these in the Half Breadth on their respective stations with the edge of the mould, A, B, C, D, E, F, and G, fair to the sight edge just transferred from the Body. Secure them, perpendicular, by a light cross piece tacked at the bottom to the floor, and attach them to each other by a fore and aft piece, P. The top outside edges of these templates give the *correct* form of the longitudinal surface and the position of the frames. A mould, like Fig. 143, is then made for each plate showing the extreme edges,

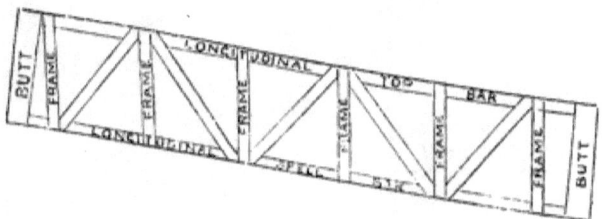

Fig. 143.

with cross pieces for the position of the frames, and end pieces for the butt straps. The latter should be arranged for the strap to clear the vertical bars. This is the method generally adopted where there is considerable twist, and absolute correctness is required. The watertight longitudinal butts are usually treble riveted and the non-watertight double.

NOTE.—It may be noted that an easy way of transferring a line from the loft floor on to a board for making a template is to place small flat-headed nails with the edge of the head resting on the line, and then lay a board on the top, and a little pressure will cause them to adhere to the board, when it may be lifted and the curve drawn in pencil through the points.

Fig. 144.

BATTLE SHIP'S BODY PLAN.

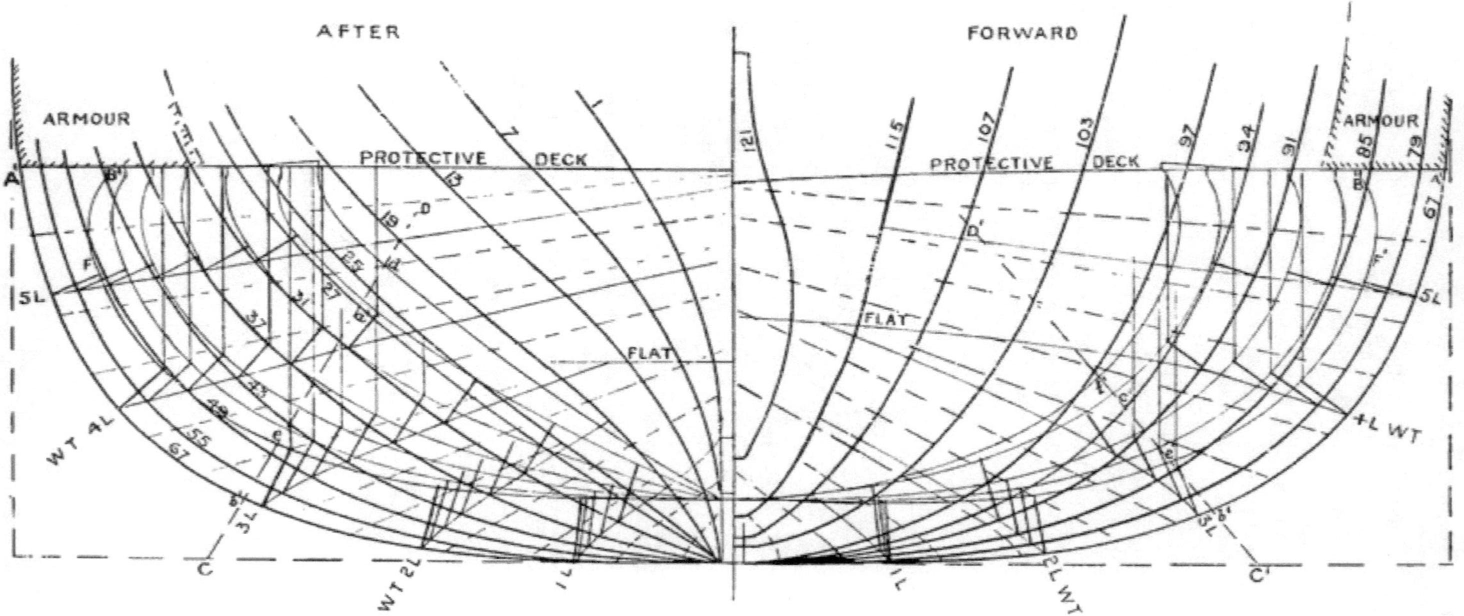

To Obtain and Fair-up the Inner Bottom Lines of a Battle Ship.—The inner bottom is different in that it extends to the armour shelf on each side. The form is first approximated on the Body sections (**Fig. 144**) in the following manner. When the ship has a raised keel, lift from the "keelson plan" the height of the centre keelson above the base line, at each section, for the length of the double bottom, and set the distances up the centre of the Body above the corresponding base. The moulded midship depth will be from 36 to 40 inches, and it may, or may not, be parallel to the keel fore and aft. Where the ship is on a level keel, with a parallel keelson, the top will be represented by one point as in the **Fig. 144**. The central part of the inner bottom is made level fore and aft at each side of the centre line, to take the width of the keelson top bars. The midship form may now be drawn in by setting off the breadth A B in the **Fig. 147**, and transferring it into the Body $A^1 B^1$. Then the curve is lifted from **Fig. 147**, the "scantling section," on the longitudinals to their specified moulded width, or it may be got by a gradual taper of about 4 to 6 inches from the centre point E to the top point F, and brought out to B^1 with an easy continuous curve. The midship is duplicated in both bodies. The end sections 27 and 94 are next settled in the same way, only the taper may be slightly more towards the top. Care being exercised not to take too much out of the floors for manholes, and yet to have easy access to all parts of the bottom. Diagonal lines C D and $C^1 D^1$ are drawn in each Body; and the outer bottom form, on this line, lifted on each section from C and C^1, and the points set off on the "contracted principle," as to the ship's length, above a base, from which the line a, b, c, in **Fig. 145**, is drawn. The moulded depth of the double bottom on the run of the diagonal C D and $C^1 D^1$ is lifted from the sections, $a^1 d$ on 27 frame, $b^1 e$ on 67, and $c^1 f$ on 94, and transferred into the **Fig. 145**, $a^1 d$ set above $a = a d^1$, $b^1 e$ above $b = b e^1$, $c^1 f$ above $c = c f^1$. A batten is pinned to these points d^1, e^1, f^1, and the inner bottom drawn a gradual and fair taper to the out bottom a, b, c. The distances, on each frame section, lifted and transferred into the Body on the diagonal, or diagonals, will give spots on the intermediate frames, and in conjunction with E and $A^1 B^1$, which is the same for all sections, will enable the approximate form on each section to be traced in. These lines are then faired up by a few additional diagonals and buttocks in the same manner as C D. The upper part may be faired by a close level line or two.

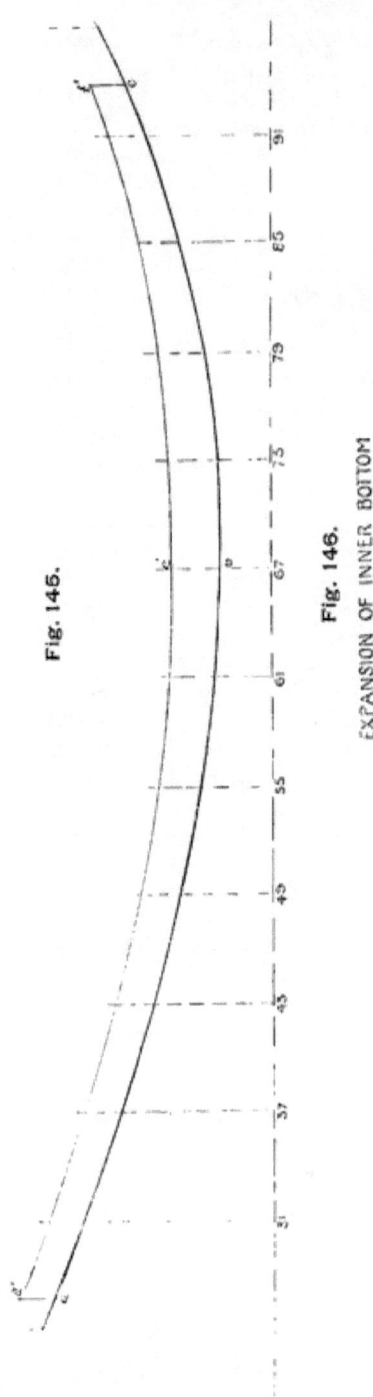

Fig. 145.

Fig. 146.

EXPANSION OF INNER BOTTOM

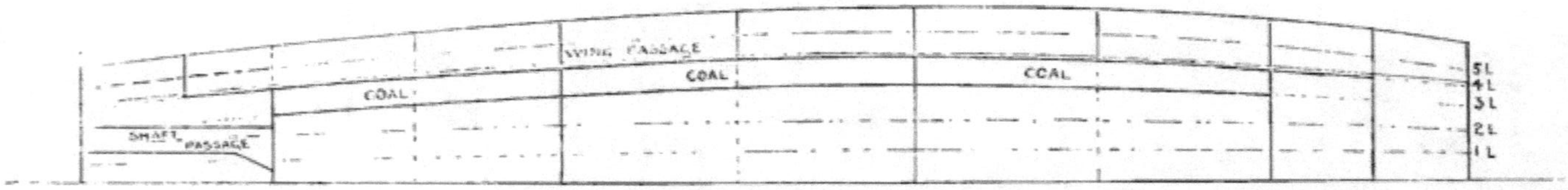

The longitudinals are then drawn square to the frames for the depth of the double bottom. It is usual to test the fairness, in the Half Breadth, of the top edge of the longitudinals in the same way as described for the outer bottom edges.

Expansion of the Inner Bottom of a Battle Ship.—An expansion of the inner bottom surface is made on a ¼-inch scale for ordering the plates, and for the guidance of the workmen in plating same.

The plating is carried to, or just beyond, the wing passage bulkhead to h, in **Fig. 147**. Girth the inner bottom, in **Fig. 144**, on each section from the centre line for the point h, and the position of the fore and aft bulkheads, longitudinals, and midship section plate edges. Lay these distances off in the expansion (**Fig. 146**) from the base or centre line. Run blue dotted lines through the points of the longitudinals and athwartship tank divisions, full blue lines showing fore and aft bulkheads, athwartship bulkheads, engine and boiler seating girders, shaft passages, and whatever will come in contact with the top of the inner bottom. Then the plate sight edges may be drawn in black ink parallel to the ship's centre, as far as they will go without crossing watertight longitudinal divisions. It may be necessary at the ends to step them back, or introduce stealers. The manholes for entrance to the inner bottom are also arranged to clear and in suitable positions. The butts of the outer bottom plating and longitudinals are indicated in red, and the butts of the inner bottom are lined off, and the plates ordered from this plan. A little allowance should be made in the length of the plates for the curve on the bottom fore and aft, or the fore and aft expanded form found.

Bevels on Inner Bottom Frames.—Where the double bottom has fore and aft curve bevels must be lifted in the usual way, commencing from the protective deck, and given on boards to the frame turners to bevel the frames.

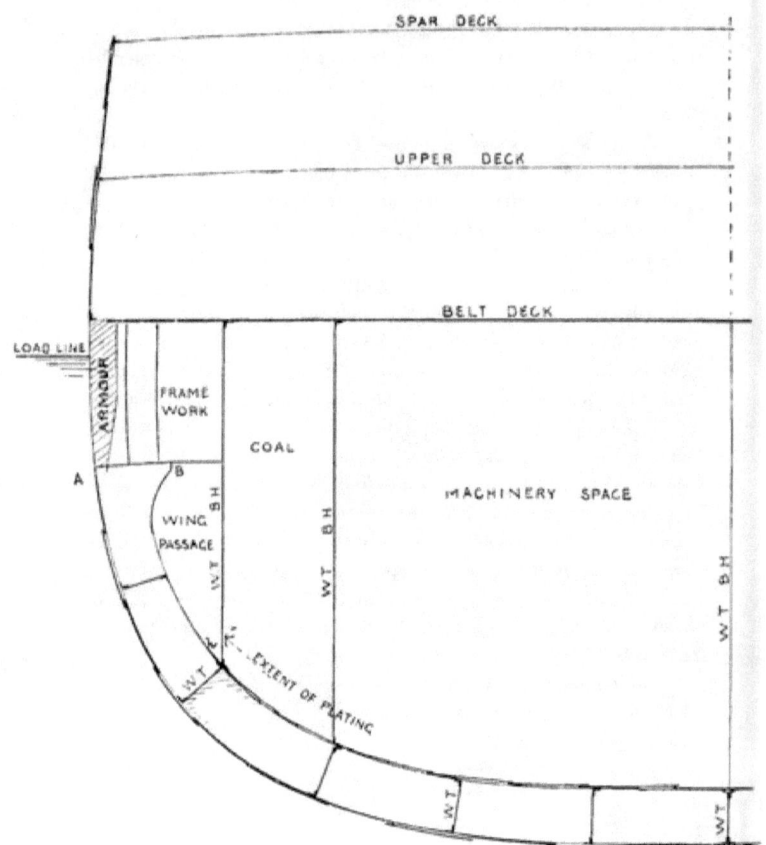

Fig. 147.

BATTLESHIP.

MIDSHIP SECTION.

CHAPTER XVI.

Obtaining and Fairing Lines of a Midship Gun Gallery—Expansion—Obtaining and Fairing Lines of a Midship Gun Gallery of Conical Type—Expansion of Conical Type—Semi Egg-shaped Forward Gun Embrasure—Another Form of End Gun Gallery—Expansion of End Gallery.

GUN GALLERIES OR SPONSONS.

It is not easy, nor necessary, to give all the different forms of galleries erected on the sides of vessels for carrying guns; for they are the outcome of the special design of ship and gun, which alters very frequently: therefore, a few of the most important and generally adopted will be described.

Midship Gun Gallery.—It is usual to draw the galleries to a large scale, say $\frac{1}{2}$ inch equal to 1 foot. For this purpose, lift the position G from the small design, and show it in **Fig. 148** with frames 60, 61, etc. Set off A B and A¹ B¹ parallel to and at a convenient and equal distance from the centre line of the vessel, and draw in the Body the form of the frame stations for the extent of the fore and aft spread, also the centre of the gallery, which lines transfer into the Half Breadth on the rail, knuckle, and at suitable levels. The form C D E, in the plan, is settled to suit the extreme range of the gun and shield. Having fixed it, lift the line C D E on each frame from A B, and lay the points out from A¹ B¹ in the Body on the knuckle level, or levels—if there is fore and aft sheer. The bulwark between the knuckle and rail is made vertical, as shown at each section. Then draw in D¹ H with a suitable angle to the ship's side, and continue the points a, b, c, d parallel to D¹ H, to meet their respective frames; projecting the cutting points into the Sheer as shown by dotted lines, through which draw the form line L M K on the ship's side. For the purpose of easy explanation, the moulded form of the ship is shown the same on both sides of the gun centre, which may not be the case. Now square down from the Sheer the cutting points of L M K with the level lines on to their corresponding levels in the Half Breadth. Lift the level line breadths from A¹ B¹ in the Body, i.e., e, f, g, h, etc., and transfer them on to their respective frames in the Half Breadth, through which, with terminations on the ship's side, draw fair form of the level lines 1 and 2. Then you have the necessary form—faired.

The bulwark part N O P, necessary for the working of the gun, may be made of hinged doors, or cut out altogether with a moulding on the outside, and an angle bar on the inside carried along the edge for stiffening.

Expansion of Midship Gallery.—It is assumed that the ship has no sheer in way of this gallery. On page 145 a method is shown for setting off the line when there is sheer.

In the expansion make $D^2 D^3$ equal to $D^1 D^4$. Draw $C^1 E^1$ through D^3 and parallel to $C^2 E^2$, and both at right angles to $D^1 D^2$. Girth rail and knuckle in the Half Breadth from D to butt C and E, marking at the same time the position of the frames. Lay the knuckle girth off about D^2, and that of the rail on D^3. Join top and bottom points, and you have expanded bulwark plating.

The lower portion, or sponson, is expanded by drawing in the Body the set line R S square to H D^1, and $R^1 R^2$ in the expansion of parallel distance $D^2 S^1$ equal to $D^1 S$. Then lift from R, in the Body, $R a^1$, $R b^1$, $R c^1$, $R d^1$, and R S, and place them on their respective sections in the expansion, which will give dotted line T T, the form of the gallery through S R. Girth this line for the position of the frames, which set-off on the set line $R^1 R^2$ around S^1. From these points the expanded frames are shown dotted in and numbered. Then measure the distance from the set line R S in the Body to the knuckle points a, b, c, d, and D^1, and lay them off on the expanded frames above $R^1 R^2$, through which points a^2, b^2, c^2, etc., draw curve. Next lift from the Body the distances, on the run of the section, below the set line R S to frame cutting points, and measure off below $R^1 R^2$ in the expansion. Trace fair curve through the spots. This gives the expanded lower edge. The figure $R^1 X R^2 D^2$ is the expanded form. The plating may be ordered in two or three plates with vertical butts; allowance being made for the knuckle and shell connection over and above this expansion.

In the case of a plate edge S W, shown in the Body, running fore and aft. Lift the distance from R S, on each section, and set-off in the expansion below $R^1 R^2$. The termination may be got by lifting cutting point at W square to R S, and measuring it off parallel to $R^1 R^2$ to cut line R^1, X, R^2, by which you get W^1 and W^2. Draw curve $W^1 S^2 W^2$ through points, and you have the line on the expansion.

Midship Gun Gallery of a Conical Type.—In **Fig.** 149, draw A B and $A^1 B^1$ in plan and cross section at a suitable, equal, and parallel distance to the centre line of the ship. Set off from $A^1 B^1$ on

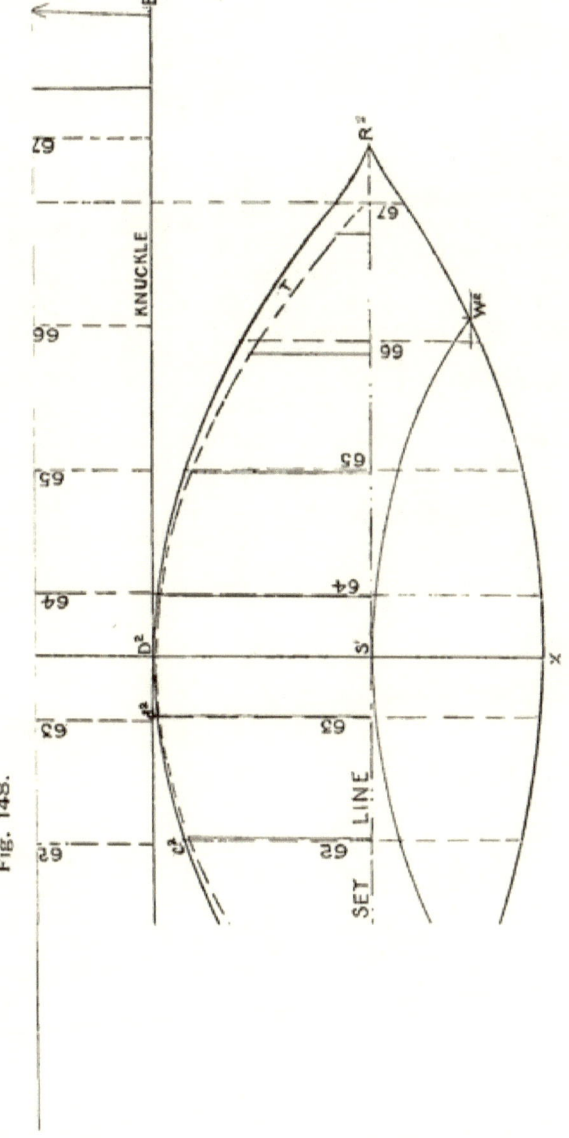

Fig. 148.

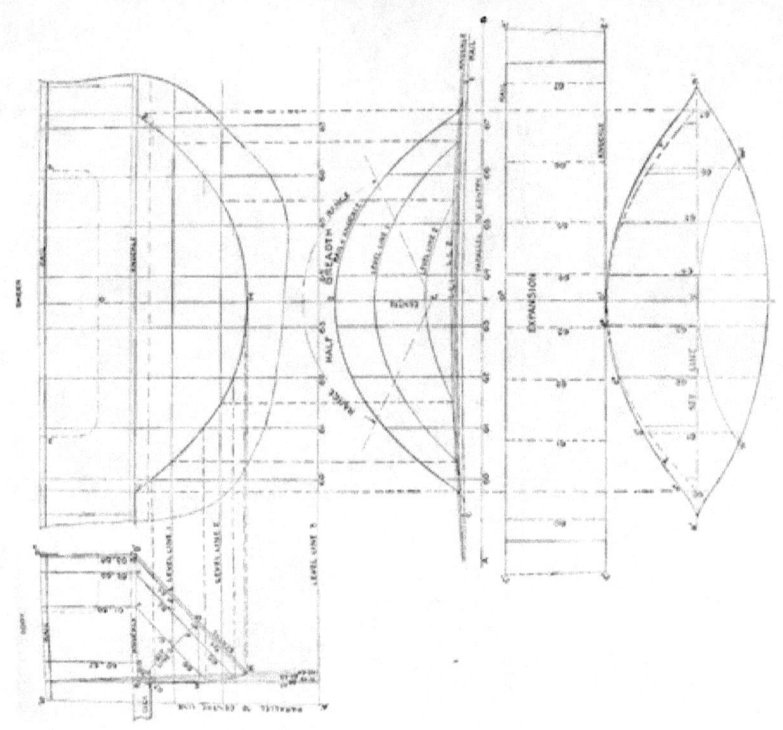

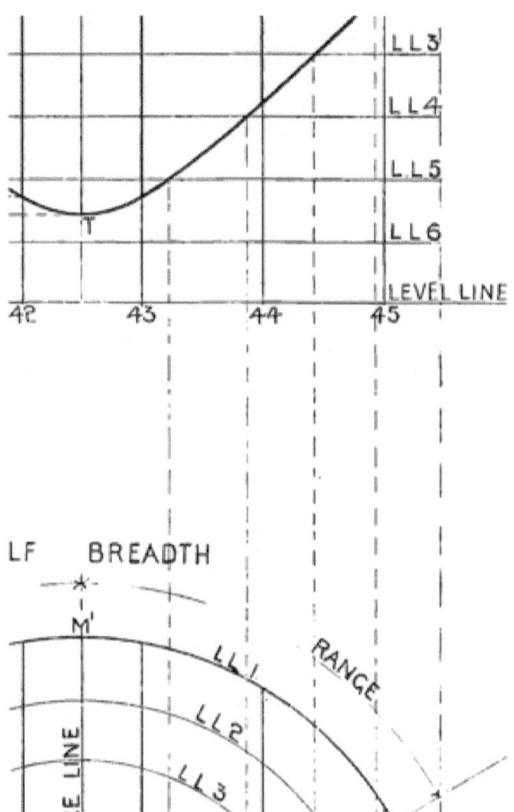

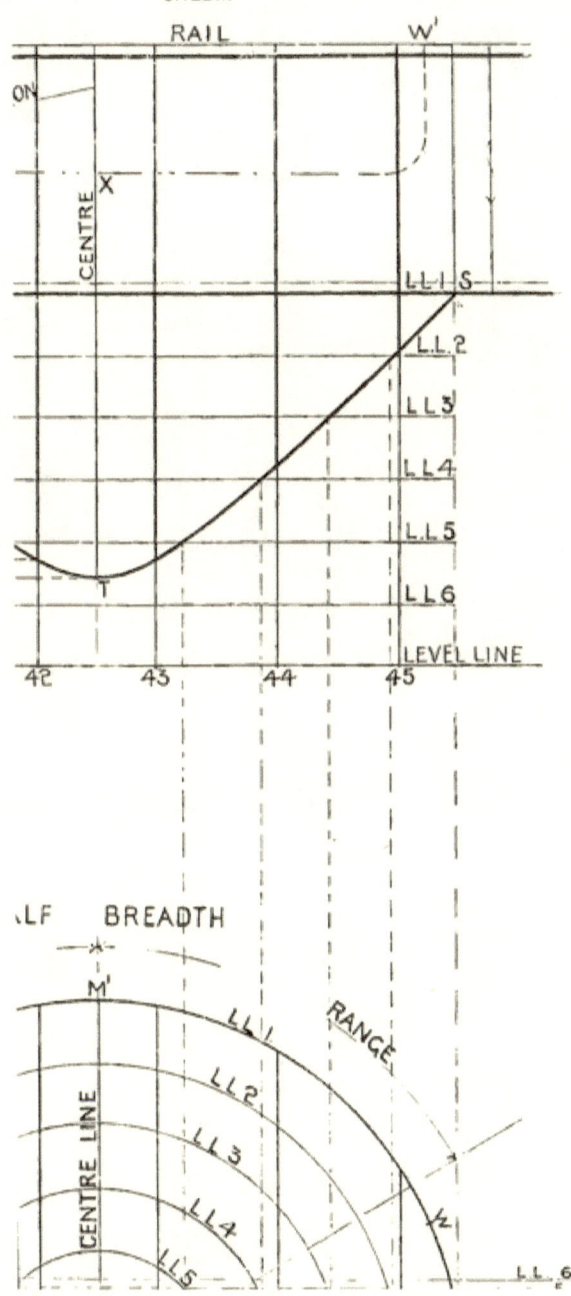

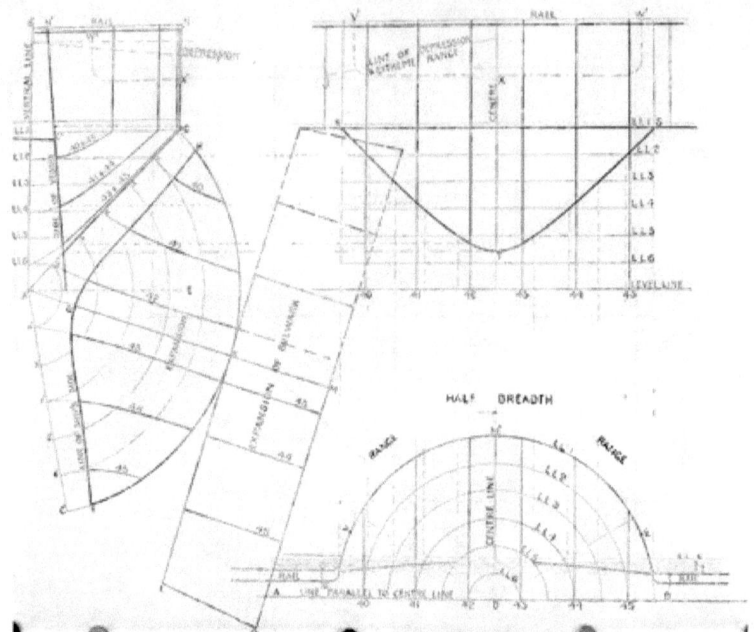

GUN GALLERIES OR SPONSONS. 143

L L 1 the outstretch of the gallery sufficient to work gun to the range and depression. Join C to A¹ with a suitable angle. The point C is generally kept on a level with the lower edge of the gunwale moulding. Draw in the Body the level lines L L 2, L L 3, etc., and produce them into the Sheer. Then, with radii equal at each level line of the Body, to the distance from the vertical A¹ B¹ to C A¹, describe in the Half Breadth from D radiating level lines, 1, 2, 3, 4, 5, and 6. Show in the Half Breadth position of the frame stations 40, 41, 42, 43, 44, and 45. Lift the ordinates of these radiating lines from A B at each station, and transfer them into the Body from A¹ B¹ on their corresponding levels. The points of termination on A¹ B¹ may be got by lifting the distances of the frame stations from D, say D to 42, D to 41, D to 40, and setting them off from A¹ along A¹ E. Square up these points to A¹ C, and level in the intersections, as shown, on to A¹ B¹, which will enable you to draw the form of the cross sections. The side of the vessel may now be shown in the Body, if not already done, relative to A¹ B¹, and the ordinates for the same on the level lines, rail, etc., transferred into the Half Breadth, for drawing in the ship's form at the various levels. The cutting points with the radiating planes are then squared into the Sheer, indicated by dotted lines; and the frame stations, cutting ship's side form in the Body, levelled over into the Sheer on to their corresponding frames, allow the form line R T S on the ship's side to be drawn in. The bulwark above C¹ C is made to take usual sheer, and in the section is vertically parallel to A¹ B¹, and carried up to the height of the under side of the hammock berthing or rail, being joined by easy curves to the ship's form before and abaft the gallery.

Expansion of the Plating of a Conical Gallery.—From A¹ as centre describe arcs of circles, as shown, with radii equal to A¹ a, A¹ b, and so on. Then girth each level line in the Half Breadth from A B for the position of the ship's side, the frame stations, and the centre. Measure the girths from A¹ C around their corresponding level lines in the expansion, $a\ a^1$, $b\ b^1$, $c\ c^1$, etc., marking the position of the side, frames, and centre at the same time, and draw lines through the spots. The figure F, G, H, K is the form of the developed plate below the knuckle C¹ C. Allowance must be made over and above this at the edges for connection.

For the bulwarks, or the part above the knuckle, continue G K to M. Make K M equal to C N, and L K square to K M, and of length equal to the girth of M¹ L¹ in the Half Breadth. Draw M O parallel to

L, L 1 the outstretch of the gallery sufficient to work gun to the range and depression. Join C to A^1 with a suitable angle. The point C is generally kept on a level with the lower edge of the gunwale moulding. Draw in the Body the level lines L L 2, L L 3, etc., and produce them into the Sheer. Then, with radii equal at each level line of the Body, to the distance from the vertical $A^1 B^1$ to $C A^1$, describe in the Half Breadth from D radiating level lines, 1, 2, 3, 4, 5, and 6. Show in the Half Breadth position of the frame stations 40, 41, 42, 43, 44, and 45. Lift the ordinates of these radiating lines from A B at each station, and transfer them into the Body from $A^1 B^1$ on their corresponding levels. The points of termination on $A^1 B^1$ may be got by lifting the distances of the frame stations from D, say D to 42, D to 41, D to 40, and setting them off from A^1 along $A^1 E$. Square up these points to $A^1 C$, and level in the intersections, as shown, on to $A^1 B^1$, which will enable you to draw the form of the cross sections. The side of the vessel may now be shown in the Body, if not already done, relative to $A^1 B^1$, and the ordinates for the same on the level lines, rail, etc., transferred into the Half Breadth, for drawing in the ship's form at the various levels. The cutting points with the radiating planes are then squared into the Sheer, indicated by dotted lines; and the frame stations, cutting ship's side form in the Body, levelled over into the Sheer on to their corresponding frames, allow the form line R T S on the ship's side to be drawn in. The bulwark above $C^1 C$ is made to take usual sheer, and in the section is vertically parallel to $A^1 B^1$, and carried up to the height of the under side of the hammock berthing or rail, being joined by easy curves to the ship's form before and abaft the gallery.

Expansion of the Plating of a Conical Gallery.—From A^1 as centre describe arcs of circles, as shown, with radii equal to $A^1 a$, $A^1 b$, and so on. Then girth each level line in the Half Breadth from A B for the position of the ship's side, the frame stations, and the centre. Measure the girths from $A^1 C$ around their corresponding level lines in the expansion, $a a^1$, $b b^1$, $c c^1$, etc., marking the position of the side, frames, and centre at the same time, and draw lines through the spots. The figure F, G, H, K is the form of the developed plate below the knuckle $C^1 C$. Allowance must be made over and above this at the edges for connection.

For the bulwarks, or the part above the knuckle, continue G K to M. Make K M equal to C N, and L K square to K M, and of length equal to the girth of $M^1 L^1$ in the Half Breadth. Draw M O parallel to

K L, and girth $M^1 O^1$ for length of M O. Connect O to L. The other side of the figure is got in the same manner. The depth of L O will be more if $C^2 N^1$ is not level with C N. In other words, account must be taken of any sheer in R S. For simplicity it is assumed that there is no sheer in this case. The gallery is sometimes made with hinged doors from V W, but in a number of cases the bulwarks are cut down as shown by dotted lines, V^1 X W^1 and W^2 X^1, to allow for firing of the gun at extreme angles. In such a case the rail moulding is carried round V^1 X W^2 with an angle bar on the inside to stiffen the edge.

The plating of the part below the knuckle may be arranged with a seam fore and aft between A^1 and C, or vertical plates with butts to clear frames may be fitted. The sponson is stiffened by angle iron frames with solid floors lightened by manholes.

Semi Egg-shaped Forward Gun Embrasure.— Occasionally, it is constructed in the following manner. In **Fig. 150**, line in the Body the moulded form of the ship at each frame in way of the gun, and place the outline in the Sheer and Half Breadth. Fix the centre A B and A^1 B^1, and trace in the gun outline with shield C. Describe, in section, the arc D E F to clear shield when at depression or elevation. Lift the distance A^1 E, from the Body, on level line 3 into the Half Breadth A E^1, and sketch in approximate horizontal form F E^1 G, such to allow the working of the gun when at extreme range. Square up F^1 and G into the Sheer on L L 3, and level over from the Body D and F on to 96 frame, by which four points are secured, and the approximate form on the ship's side is drawn. Line off in the Sheer and Body closely spaced level lines, and square the cutting points in the Sheer into the Half Breadth on to their corresponding level lines, which, in conjunction with the distances of J, H, K, and L lifted from Body on to A E^1, will allow the approximate form of the level lines in the Half Breadth to be drawn. Transfer this form at each section into the Body; and fair-up in the Half Breadth on diagonal lines like M N and M O. The form of the embrasure should be so arranged that the gun will clear the plating in all possible positions. A port P R S is made for working the gun at extreme angles, which means, that the lower edge sill should clear when at a depression of say 25 degrees, and the top edge at an elevation of 30 degrees at 80 degrees range. Doors are fitted to this port, hinged fore and aft, or towards the forecastle deck. The plating, forming the embrasure, is usually worked in two, butted flush as shown, and edges flanged on to the ship's side. The form cannot well be developed, so that the

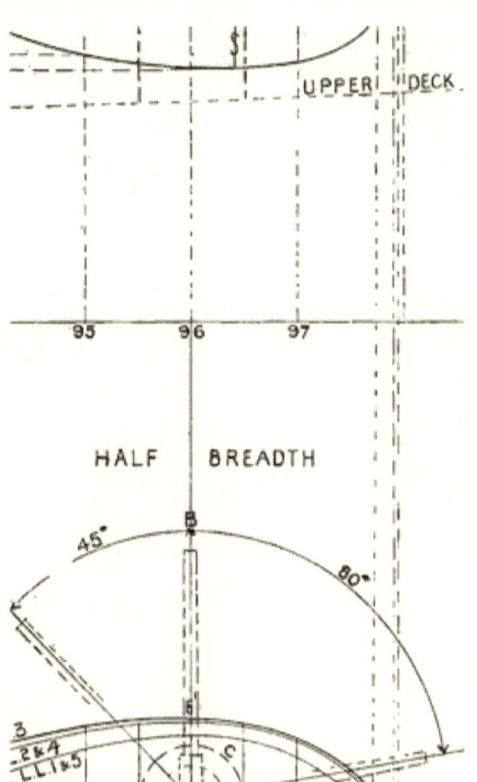

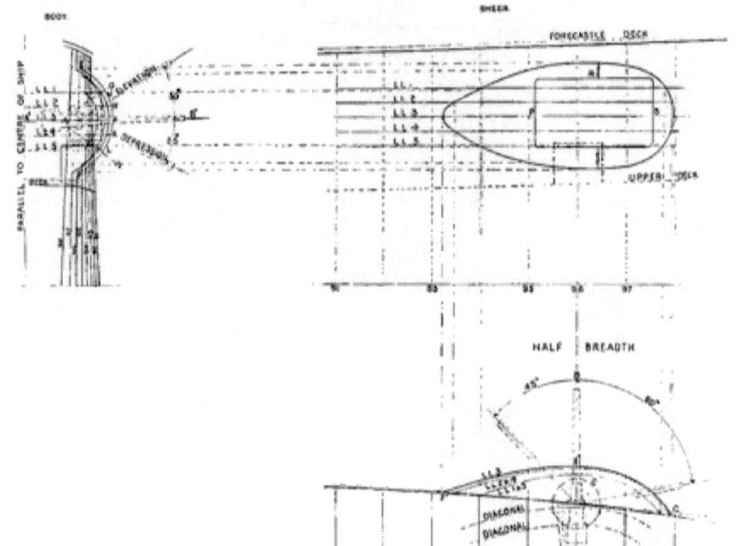

Fig. 150.

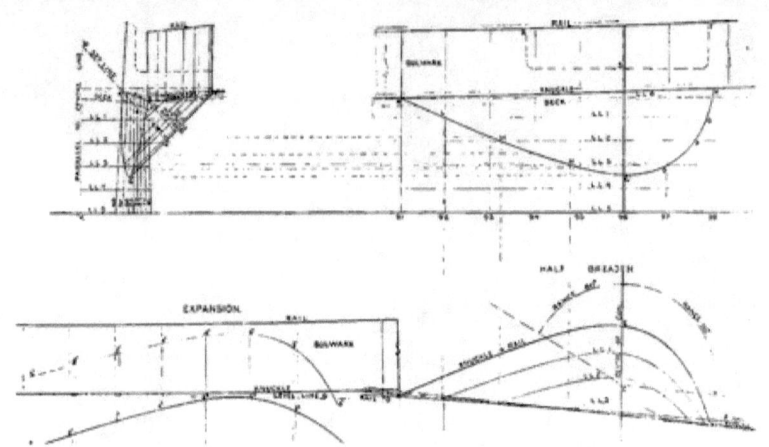

loftsman fits up section moulds on the ship at each frame, from which the platers make their own templates. This form may also be arranged at the aft end in the same manner.*

Another form of End Gun Gallery—sometimes fitted. — Referring to Fig. 151, make A B and A¹ B¹ of equal parallel distance to the centre line of the ship, and draw in the Body the moulded form of the ship's sections above the load-line relative to A¹ B¹. Lift these into the Half Breadth and Sheer, on convenient level lines, knuckle, and rail. Produce the level lines into the Sheer. Fix gun position C and the range fore and aft, make D E F a fair curve suitable to work gun at range and depression. Level knuckle sheers, on the frames, into the Body, and lift the distance of D E F from A B on each frame, and set them off on their corresponding sheer levels in the Body. Make the centre section E¹ G of suitable angle to ship's side—the point G is kept well above the load-line—and produce the sheer knuckle points parallel to the line E¹ G to meet respective frames of the ship's side. Then, project these cutting points on the ship's side into the Sheer on to their corresponding frame stations, and run line H G¹ K, which is the form line in the Sheer. Fair-up, by transferring the cutting points on each level line from A¹ B¹ of the Body to the Half Breadth, setting them out on their stations from A B. The terminations will be got by squaring down from the Sheer on to the Half Breadth corresponding lines H, L, M, N, O, P, Q, and K, which will allow the level lines to be drawn in. The bulwark is made perpendicular, as shown; and, usually, of parallel sheer to the knuckle. It is fitted with hinged doors for working gun, or it may be cut away in the manner of the dotted line R S T to suit range and depression, and stiffened on its edge with angle bar and half-round moulding.

Expansion of End Gun Gallery. — Show in the Body the set line V B¹ square to E¹ G cutting point a. Lift from B¹ on B¹ V, points a, b, c, d, e, f, and g, and lay them off on the expansion frames, (ordinary spacing) on the line A B produced, V² B² ; through the points a^1, b^1, c^1, etc., draw in $h^1 a^1$. Girth this line around c^1, station 96, for the frame points d^1, e^1, f^1, g^1, and h^1, and lay them off abaft of 96 on V² B², and also b^1 and a^1 forward of 96 in the same manner. Erect perpendiculars through spots (dotted lines), which are the expanded section positions. Lift the knuckle points, in the Body, *on* the stations above the set line B¹ V, and place them on the expanded

* It is usual to give 90 degrees range forward of A B, so that direct fore and aft fire may be secured.

loftsman fits up section moulds on the ship at each frame, from which the platers make their own templates. This form may also be arranged at the aft end in the same manner.*

Another form of End Gun Gallery—sometimes fitted.—Referring to Fig. 151, make A B and A¹ B¹ of equal parallel distance to the centre line of the ship, and draw in the Body the moulded form of the ship's sections above the load-line relative to A¹ B¹. Lift these into the Half Breadth and Sheer, on convenient level lines, knuckle, and rail. Produce the level lines into the Sheer. Fix gun position C and the range fore and aft, make D E F a fair curve suitable to work gun at range and depression. Level knuckle sheers, on the frames, into the Body, and lift the distance of D E F from A B on each frame, and set them off on their corresponding sheer levels in the Body. Make the centre section E¹ G of suitable angle to ship's side—the point G is kept well above the load-line—and produce the sheer knuckle points parallel to the line E¹ G to meet respective frames of the ship's side. Then, project these cutting points on the ship's side into the Sheer on to their corresponding frame stations, and run line H G¹ K, which is the form line in the Sheer. Fair-up, by transferring the cutting points on each level line from A¹ B¹ of the Body to the Half Breadth, setting them out on their stations from A B. The terminations will be got by squaring down from the Sheer on to the Half Breadth corresponding lines H, L, M, N, O, P, Q, and K, which will allow the level lines to be drawn in. The bulwark is made perpendicular, as shown; and, usually, of parallel sheer to the knuckle. It is fitted with hinged doors for working gun, or it may be cut away in the manner of the dotted line R S T to suit range and depression, and stiffened on its edge with angle bar and half-round moulding.

Expansion of End Gun Gallery.—Show in the Body the set line V B¹ square to E¹ G cutting point a. Lift from B¹ on B¹ V, points a, b, c, d, e, f, and g, and lay them off on the expansion frames, (ordinary spacing) on the line A B produced, V² B² ; through the points a^1, b^1, c^1, etc., draw in h^1 a^1. Girth this line around c^1, station 96, for the frame points d^1, e^1, f^1, g^1, and h^1, and lay them off abaft of 96 on V² B², and also b^1 and a^1 forward of 96 in the same manner. Erect perpendiculars through spots (dotted lines), which are the expanded section positions. Lift the knuckle points, in the Body, on the stations above the set line B¹ V, and place them on the expanded

* It is usual to give 90 degrees range forward of A B, so that direct fore and aft fire may be secured

11

frames above the set line $V^2 B^2$, viz., a^2, b^2, c^2, etc. The point b^2 may be got by lifting $o\,p$ square off from the line $B^1 V$, and setting it out from $B^2 V^2$. Then lift the distances *on* the stations below $B^1 V$ to ship's side, and lay them off below the set line in the expansion, through which points draw the curve a^2, c^2, and b^2. This figure, b^2, c^2, a^2, c^2, is the expanded form below the knuckle; allowance must be made in ordering for riveting to the ship's side and bulwarks.

The bulwark plating may be got by girthing the positions of the frames and butts D and F on the knuckle line D E F, and laying them off in the expansion on the level line 6, about 96 frame, from which erect frame perpendiculars, and on them set off above level line the sheer of the knuckle on each frame. A curve passed through these spots will give expansion of the knuckle line, and the rail may be drawn parallel to this, or it may be laid off independently in the same manner. The butt on the rail edge is got by girthing its position in the Half Breadth on each side of E, and laying it off about 96 on the rail expansion line. It may be necessary to get the true form of the butt to place in a level line between the rail and knuckle.

CHAPTER XVII.

Principal Moulds and the Order they are sent into the Yard—Stern Posts—Stems—Stern Tubes—Struts—Beam Camber—Conning Tower—Pilot Bridge—Boat Davits and Chocks.

MOULDS.

After the ship is faired-up on the loft floor wood moulds, or templates, are supplied for constructing many of the parts. Owing to the complexity and close accuracy required in war ships they are more numerous than in merchant work. It is scarcely possible to state all the moulds, for every vessel will create its own necessity; but the following are some of the principal named in the order of sending out.

1. Flat keel plate.
2. Vertical centre keelson.
3. Stem and stern posts.
4. Stern tubes and struts in twin screws.
5. Beam camber.
6. Longitudinals and armour shelf.
7. Protective deck and stringers.
8. Backing plating.
9. Belt or side armour.
10. Belt deck.
11. Barbette or redoubts.
12. Gun galleries.
13. Conning tower.
14. Pilot bridge.
15. Boat davits and chocks.

Nos. 1, 2, 5, and 6 to 12, are explained on pages 79, 83, 34, 130, 117, 119, 124 and 141. The remainder will now be briefly described.

Stern Posts are made of solid or hollow section and of cast steel. It is outside of the province of this treatise to give the different forms of sterns which are adopted. The object being simply to show what is expected from the loft for manufacturing purposes; therefore, only one form of post is given, **Fig. 152**, that of a vessel with twin screws. The figure shows a longitudinal sectional elevation drawn by the loftsman on loose boards, two or more, with sections at various points C D, A B, K L, E F, H J, and at the frame stations 1, 2, 3, etc. In some cases a rabbet, single or double, is made in the casting to receive the shell plating, sufficiently deep to make it, when in position, flush with the extreme side of post. Where the fore and aft bottom piece is of considerable length an extra stiffening marked G is cast on the post—the outline is shown by dotted lines in the figure. The shell plating is attached to this stiffener. It will be noted that the sketch shows a double notch for the flat keel plates; this, in some cases, is only single where a single plate keel is fitted. It is usual to indicate on the moulds the position of the level or water-lines. These boards are sent to the steel works, or yard pattern shop, where suitable patterns are made from them for casting the post.

Stems are also made of solid or hollow section and of cast steel. **Fig. 153** is a full-sized drawing made by the loftsman on loose boards, showing the outside and inner edges and rabbet-line, etc. It is customary to cast this form of stem in two pieces with a V keyed scarph, placed on or about the load water-line. The drawing gives the necessary information required for making the patterns. A single notch is shown for the flat plate keel. In the case of a double plate keel another notch should be made. War vessels are generally fitted with flat plate keels.

Stern Tubes.—Almost all the large vessels are fitted with twin screws, requiring watertight tubes built into the structure where the shafts pass through the ship's sides. They are sometimes built in the ordinary way, with plates and frame bars; the frames being bossed in and out to suit the tube and a cast steel narrow ring fitted at each end to take the engineer's brass tube. In other cases a steel casting is fitted the full length, with ends formed as in **Fig. 154**, and the part between F and K reduced in thickness to about $1\frac{1}{2}$ inches similar to **Fig. 166**. **Fig. 154** represents another method of construction. The sectioned ends are of cast steel of at least $1\frac{1}{2}$ inches thick at any part. A complete built steel tube of $\frac{1}{2}$ to $\frac{5}{8}$ of an inch thick is placed between the end castings, connected by a lap joint; a rabbet being formed for the purpose. Cross sections are given at various points showing how the shell plating is attached to the castings and tube. At A, N, J, and L partial bulkheads are fitted, to which palms from the castings are riveted. At the after end there is a projecting ledge, with a notch to take the shell plating. The frames O, G, and

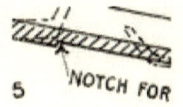

5 NOTCH FOR

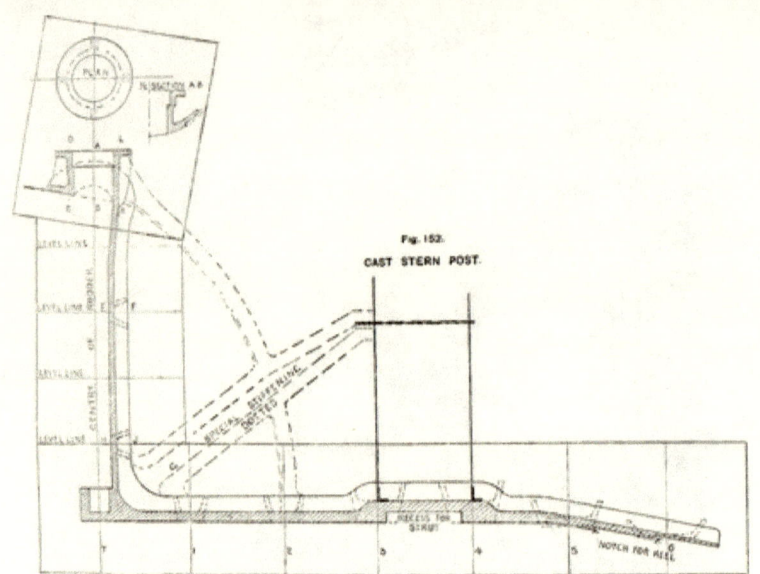

Fig. 152.
CAST STERN POST.

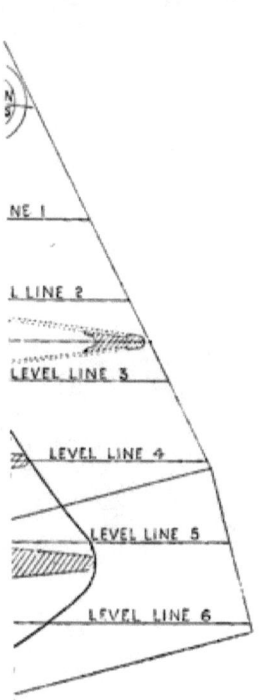

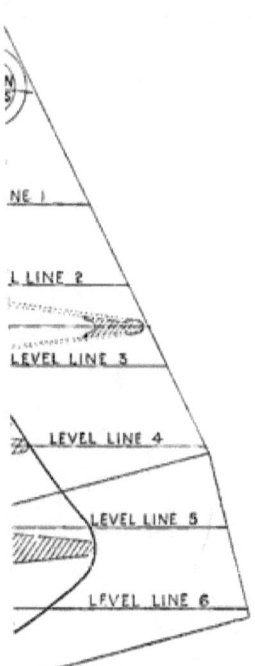

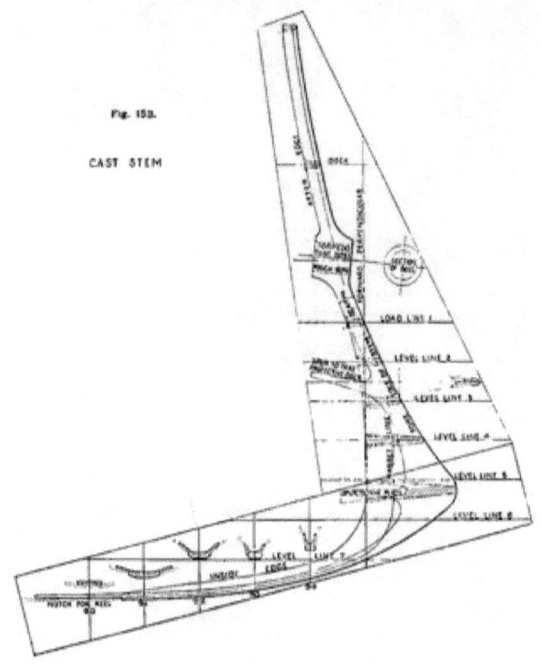

Fig. 153.

CAST STEM

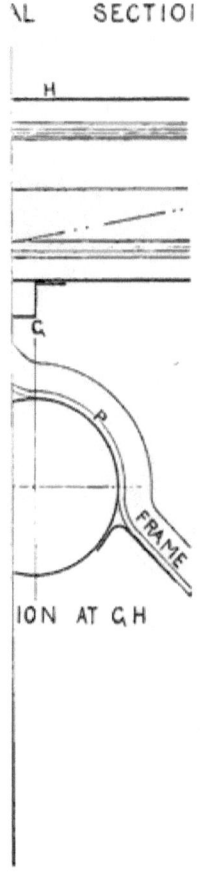

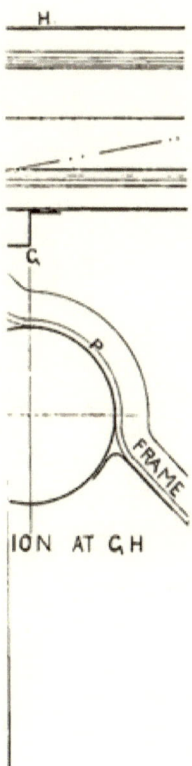

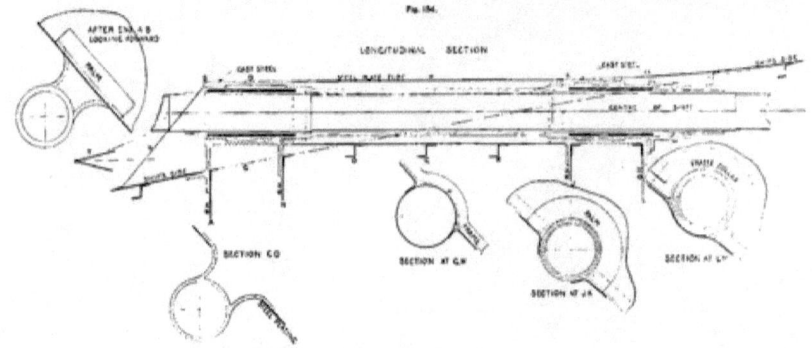

R are bossed in to the shape of the tube, and the two riveted together about P. One or more frames are cut at the fore end to allow the withdrawing of the engineer's tube and stuffing box: fore-and-afters are fixed above and below to maintain the strength. Just abaft of A B two light plates S are placed around shaft to ease the corner. They are riveted to the shell and A B, having a tapering form from A B to the aftermost edge, like T section. The space behind the plates may be filled in with light wood.

The loftsman supplies the pattern-makers, for casting purposes, with a full-sized drawing of the sectioned parts, taken, as far as possible, square off: and sections at A B, C D, N, E F, and J K, U V, L M, having winding vertical and horizontal lines to enable them to secure the proper bevel of the vessel's side.

Struts or After End Supports.—In twin screws the after ends of the shafts are supported by projecting arms from the ship's side, the same as those given on page 162 for sheathed vessels. The upper palms are commonly attached to the protective deck plating, or a special saddle plate fitted for the purpose.

Beam Camber.—Beam moulds, as explained on page 34, are supplied for all decks, only the beam arms are made to suit the special requirements of the structure, and in many cases for a double vertical row of rivets. The camber given to beams is not quite so much as that required by the Classification Registry.

Conning Tower.—At least one is fitted at the fore end on the upper or spar deck just abaft of the barbette, formed as shown in sectional plan and elevation, **Fig. 155**. The armour plating varies from 4 to 10 inches in thickness, and is supported on a rigid raised platform attached to the deck. The after end, as seen in the figure, has an open entrance of about 20 inches, protected by an overlapping screen of slightly thinner armour. The vertical edges of the screen and entrance are rounded away. The main part is formed of two plates with a vertical butt at the centre line, attached together by a vertical key. Sometimes the armour is carried up to the crown plate and attached by strong bars. This crown plate is about 2 inches thick. The armour may be kept short of the top to get a thinner plate between it and the crown to take the sight holes. These holes are 24 inches long horizontally and 3 inches deep vertically: the centre being level with the eye of an average man when standing.

In ordering the armour plating, a full-sized plan showing the butt and key-way, is scrieved in on a board by the loftsman. This is

R are bossed in to the shape of the tube, and the two riveted together about P. One or more frames are cut at the fore end to allow the withdrawing of the engineer's tube and stuffing box; fore-and-afters are fixed above and below to maintain the strength. Just abaft of A B two light plates S are placed around shaft to ease the corner. They are riveted to the shell and A B, having a tapering form from A B to the aftermost edge, like T section. The space behind the plates may be filled in with light wood.

The loftsman supplies the pattern-makers, for casting purposes, with a full-sized drawing of the sectioned parts, taken, as far as possible, square off; and sections at A B, C D, N, E F, and J K, U V, L M, having winding vertical and horizontal lines to enable them to secure the proper bevel of the vessel's side.

Struts or After End Supports.—In twin screws the after ends of the shafts are supported by projecting arms from the ship's side, the same as those given on page 162 for sheathed vessels. The upper palms are commonly attached to the protective deck plating, or a special saddle plate fitted for the purpose.

Beam Camber.—Beam moulds, as explained on page 34, are supplied for all decks, only the beam arms are made to suit the special requirements of the structure, and in many cases for a double vertical row of rivets. The camber given to beams is not quite so much as that required by the Classification Registry.

Conning Tower.—At least one is fitted at the fore end on the upper or spar deck just abaft of the barbette, formed as shown in sectional plan and elevation, Fig. 155. The armour plating varies from 4 to 10 inches in thickness, and is supported on a rigid raised platform attached to the deck. The after end, as seen in the figure, has an open entrance of about 20 inches, protected by an overlapping screen of slightly thinner armour. The vertical edges of the screen and entrance are rounded away. The main part is formed of two plates with a vertical butt at the centre line, attached together by a vertical key. Sometimes the armour is carried up to the crown plate and attached by strong bars. This crown plate is about 2 inches thick. The armour may be kept short of the top to get a thinner plate between it and the crown to take the sight holes. These holes are 24 inches long horizontally and 3 inches deep vertically; the centre being level with the eye of an average man when standing.

In ordering the armour plating, a full-sized plan showing the butt and key-way, is scrieved in on a board by the loftsman. This is

Fig. 155.

CONNING TOWER

FORE & AFT ELEVATION

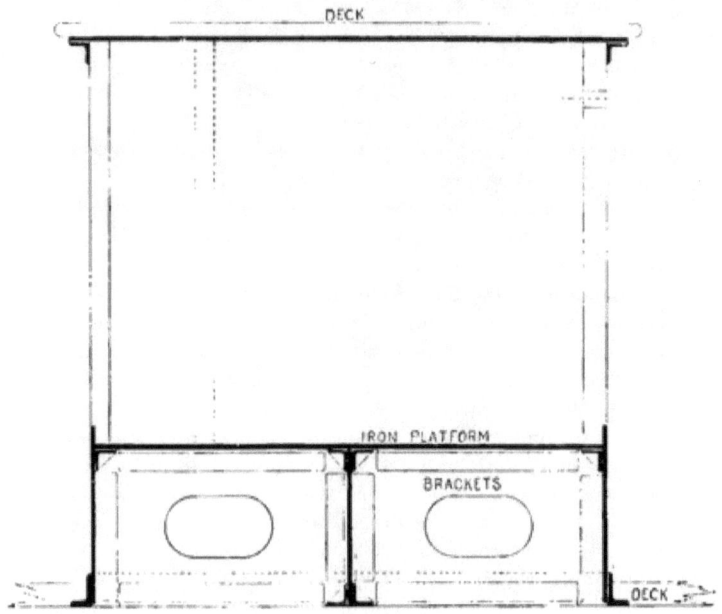

PLAN

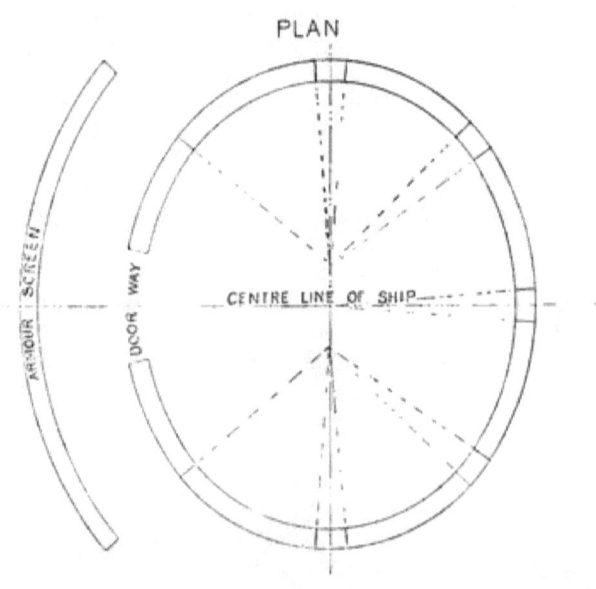

sent with the finished depth to the manufacturers. Where the armour is carried to the crown plate, the position of the sight holes should be given as in plan, **Fig. 155**, and the vertical height, with their depth, supplied to the manufacturer. A template of the seat for the armour is given to the yard to form platform.

Pilot Bridge.—This is laid down on the loft floor and a wood skeleton template, for the use of the yard, is made to the heel of the foundation bar for one side of the ship only. If there is much camber, the heel line should be expanded and a camber mould supplied with the template.

Boat Davits and Chocks.—Occasionally a flat wood mould showing the edges and head through the vertical centre is supplied the forge, having painted on at various points of the length the required cross section. It is usual in this case to lay down on the loft floor the plan form of the gunwale and the half sectional form of the boat at the davit points. Then to line in the edges of the davits for making the moulds, in such a manner that there will be room for fenders, and to get the boat in and out of the davits, and to clear the ship's rail and side. The keel of the boat is generally arranged when swinging out, 12 inches clear of the rail ; and, the broadest beam, 6 inches clear of the ship's side when waterborne at the load-line.

Chock moulds to the inside edge are made for all boats for the use of the forge or angle smiths, due allowance being made for padding. These should be accompanied with fore and aft bevels to suit the form of the bottom : the points chosen are indicated on the moulds.

In many yards a figured drawing is supplied of the davits, showing the height, outstretch, radius, and diameter at different parts, with an enlarged figured detail sketch of the head, which is considered ample for their construction.

CHAPTER XVIII.

To Obtain and Line off the Draught Marks on Stem and Stern.

DRAUGHT MARKS.

Draught Marks.—These are placed on both sides of the stem and stern, by means of which the draught of water that the ship draws at any time when afloat may be correctly ascertained. In some cases the metrical system is applied on one side and the English figures on the other. **Fig. 156** shows the method of lining off these marks. Select a line A B, square to the load-line, and as near the perpendicular as possible. Then the bottom side of the keel F is produced to C. On a level line with A the top edge of two straight edges, about 12 by 1½ inches section, are fixed level across the ship, at a short distance apart, D and E. On the top of these are set and fixed square to load-line, as shown in half cross section, two long measuring laths, about 3 feet apart; the upper part being attached to some of the staging round about. The duplicate draught marks are levelled on to the ship's side, at positions D and E, by a straight edge shown in the section at 20 feet, and each set of spots connected by chalk lines. Then the figures, which are 6 inches deep, are indicated on the side about A B, and centre-punched. Five feet is about as low as they will be required. The process is repeated on the other side of the vessel, and the after end marked off in the same way.

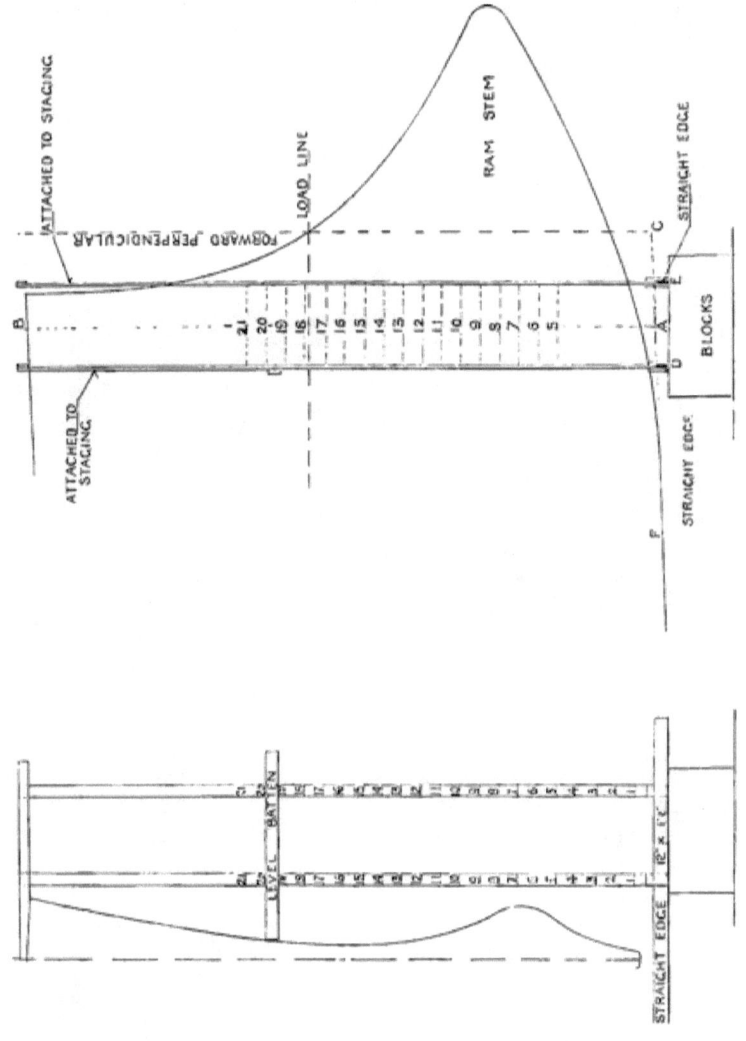

# COMPOSITE VESSELS.

CHAPTER XIX.

Sheer Draught—Extreme Form for the Calculation of Displacement—To Find the Heel of the Frames by an Approximate Method—An Exact Method of Finding the Heel of the Frames—To Find the Middle Line of Rabbet—To End a Level Line in the Half Breadth—To Terminate a Frame in the Body—To Find the Bearding Line approximately—To Find the Middle Line of Rabbet by an Exact Method—To Find the Bearding Line by another Method—Form of the Rabbet in the Main Keel Piece—Working Base Line.

In composite vessels the Sheer Draught is made to the outside of the planking. This is faired-up first on the loft floor, and the displacement checked. To enable you to place the form of the frames on the scrieve board, it is necessary to remove the planking, and seeing it is made of almost uniform thickness *square* to the outside surface, with the exception of the topsides and the extreme ends, it will be evident that it is a variable quantity on the level lines, etc.

To Find the Heel of the Frames.—The method of removal of planking is simple on and about the midship section, for the frame is about square to the form of the level line in the Half Breadth. In **Fig. 157**, on the midship frame 48, set in square to the outside edge at the level lines and buttocks, the thickness of the plank $w\,o$, $z\,y$, $e\,f$, $g\,h$, and pass a curve through the points. This curve is the inside of the plank, or the heel of the frame. The process is more difficult as you approach the ends, where the level lines cut the frame stations obliquely in the plan and sections. A quick method is to set in square to the outside edge at the several level lines and stations in the Half Breadth the thickness of the plank $s\,t$ on say 90 frame, and transfer the distance $r\,s$ into the body $s^1\,r^1$, square to the section. By this means a sufficiency of spots is got to draw in S T, the inside of the plank. Then the cutting points *on* the level and buttock lines are lifted into the Half Breadth and Sheer respectively, and the inside form faired-up in the usual way.

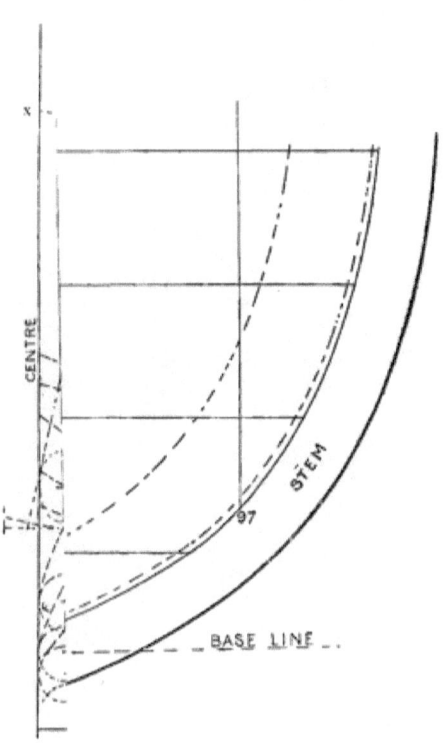

FULL LINES—OUTSIDE OF PLANKING.
DOTTED LINES—INSIDE OF PLANKING.

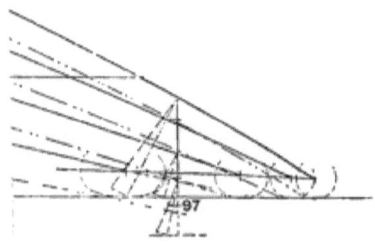

Fig. 157.

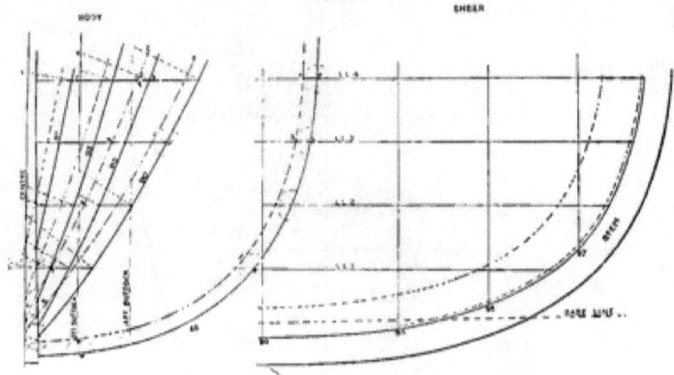

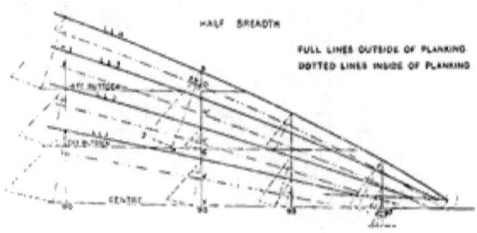

FULL LINES OUTSIDE OF PLANKING
DOTTED LINES INSIDE OF PLANKING

A More Exact Method of Finding the Heel of the Frames.
—This is given in Professor Rankine's *Shipbuilding, Theoretical and Practical.* It is the one which has been adopted in large shipbuilding establishments for many years.

In **Fig. 157**, 4 L L is a level line showing the outside of the planking.

At the point B and *b* erect perpendiculars B A and *b a* to 4 L L and 90 frame respectively. On X B set off any distance X C, and make *x c* equal X C. Through *c* and C draw *c a* and C A perpendicular to *b x* and B X, cutting *b a* and B A at *a* and A respectively. From A draw A D perpendicular to A B, and make A D equal to *a c*. Join B D. Make B R equal to the thickness of the plank, draw R P parallel to D A, then P is the horizontal projection of a point in the outer surface of the framing. Draw P Q parallel to A C, cutting in Q. In the body make *x q* equal X Q. Draw *q p* perpendicular to *b x*, cutting *a b* in *p*, then *p* is the vertical projection of a point on the outer surface of the frames, P being its horizontal projection. This point is not situated either in the plane of the level line or that of the square station; it is therefore necessary to find similar points as shown in the **Fig. 157** for each frame station, then a curve E F may be passed through these, and its intersection k, n, m, l, with the plane of each level line, will give a point through which the level line will pass. The distances k, n, m, l, on each section are measured off and transferred into the Half Breadth on their corresponding stations, and curves passed through the points k^1, n^1, m^1, l^1 will show the form of the inner surface on the level lines. The buttock heights may also be lifted to the dotted or inner surface and transferred into the Sheer and the ship faired-up on these new lines.

To Find the Middle Line of Rabbet.
—**Fig. 158** shows part of the stem and the fore end sections drawn to the plank *inside*. From the point B in the Sheer draw B D square to the fore edge of the stem. From the cutting points F, E, D, and G the outer boundary line of the plank, erect perpendiculars D H, E J, etc., of indefinite length. Level over into the Body on to their respective stations the points D, E, F, G, and lift the distances from the centre line and lay D^1 out from D on D H, E^1 from E on E J, F^1 from F on F K, and G^1 from G on G L. From L, which is the half siding of the stem at G^1, describe a circle with a radius L N equal to the required thickness of planking at that point. Draw L O square to the outside surface, then O is the middle of the rabbet, which allows the curve H, J, K,

A More Exact Method of Finding the Heel of the Frames.
—This is given in Professor Rankine's *Shipbuilding, Theoretical and Practical*. It is the one which has been adopted in large shipbuilding establishments for many years.

In **Fig. 157**, I L L is a level line showing the outside of the planking.

At the point B and b erect perpendiculars B A and b a to 4 L L and 90 frame respectively. On X B set off any distance X C, and make x c equal X C. Through c and C draw c a and C A perpendicular to b x and B X, cutting b a and B A at a and A respectively. From A draw A D perpendicular to A B, and make A D equal to a c. Join B D. Make B R equal to the thickness of the plank, draw R P parallel to D A, then P is the horizontal projection of a point in the outer surface of the framing. Draw P Q parallel to A C, cutting in Q. In the body make x q equal X Q. Draw q p perpendicular to b x, cutting a b in p, then p is the vertical projection of a point on the outer surface of the frames, P being its horizontal projection. This point is not situated either in the plane of the level line or that of the square station; it is therefore necessary to find similar points as shown in the **Fig. 157** for each frame station, then a curve E F may be passed through these, and its intersection k, n, m, l, with the plane of each level line, will give a point through which the level line will pass. The distances k, n, m, l, on each section are measured off and transferred into the Half Breadth on their corresponding stations, and curves passed through the points k^1, n^1, m^1, l^1 will show the form of the inner surface on the level lines. The buttock heights may also be lifted to the dotted or inner surface and transferred into the Sheer and the ship faired-up on these new lines.

To Find the Middle Line of Rabbet.—Fig. 158 shows part of the stem and the fore end sections drawn to the plank *inside*. From the point B in the Sheer draw B D square to the fore edge of the stem. From the cutting points F, E, D, and G the outer boundary line of the plank, erect perpendiculars D H, E J, etc., of indefinite length. Level over into the Body on to their respective stations the points D, E, F, G, and lift the distances from the centre line and lay D¹ out from D on D H, E¹ from E on E J, F¹ from F on F K, and G¹ from G on G L. From L, which is the half siding of the stem at G¹, describe a circle with a radius L N equal to the required thickness of planking at that point. Draw L O square to the outside surface, then O is the middle of the rabbet, which allows the curve H, J, K,

O, the inner surface of the plank, to be drawn in. This process is repeated at A and C, at the load-line, and sections abaft of A. A line run through the points so found will give the rabbet at any section, shown by dotted line.

To End a Level Line in the Half Breadth.—An easy way is to draw in the Sheer S T through the extreme points d, e, f, g, and h, and W V through the corresponding edge. Project down into the Half Breadth on to the centre the intersection of the outer boundary line or fore edge of the rabbet, and set off on these the half siding of the stem, through which draw parallel lines to the centre. Then square down from 1 level line the points a, c, d on to its parallel line, and mark out an ellipse through the points, the cross width will equal X x. The inner surface should terminate at b^1 on the edge of the ellipse. The same process can be repeated at each level line.

To Terminate a Frame in the Body.—Level over from the Sheer in **Fig. 158** the points of intersection m, n, p, G, of say 97 frame, with S T, W V, middle rabbet, and boundary line. Draw in an ellipse in the same manner as before described for the level line. Then 97 frame will terminate tangent to this, or the middle rabbet levelled over.

To Find the Bearding Line.—In **Fig. 158**, produce L parallel to the centre B D until it cuts the curve H K O, then make M N square to B D; N is a point on B D for the bearding line. This is done at each section until a sufficient number of spots is secured to draw line in. It may require some modification above the load-line to get a good connection for the planking.

To Find the Middle Line of Rabbet by another Method.[*]—In **Fig. 160**. Let $a\ b$ and $a^2\ b^2$ be the vertical traces of a level plane, and let $a^1\ b^1$ be the projection in the Half Breadth of its intersection with the surface of the plank. At the point b, where the level line in the Sheer cuts the fore edge of the rabbet, draw a tangent line $c\ d$ to that curve, cutting the base line of the Sheer at the point d. Project the point d upon the middle line of the Half Breadth at d^1, from which draw the perpendicular $d^1 g$, making $d^1 g =$ to $f^1 g^1$ in the Body. Then g is the horizontal trace of the tangent to the fore end of rabbet at b. Through b^1, the fore edge of the rabbet in the Half Breadth, at the height of the level line $a^1\ b^1$, draw a tangent $h\ k$ to the latter line, cutting the middle line

---

[*] Taken from Professor Rankine's *Shipbuilding, Theoretical and Practical*.

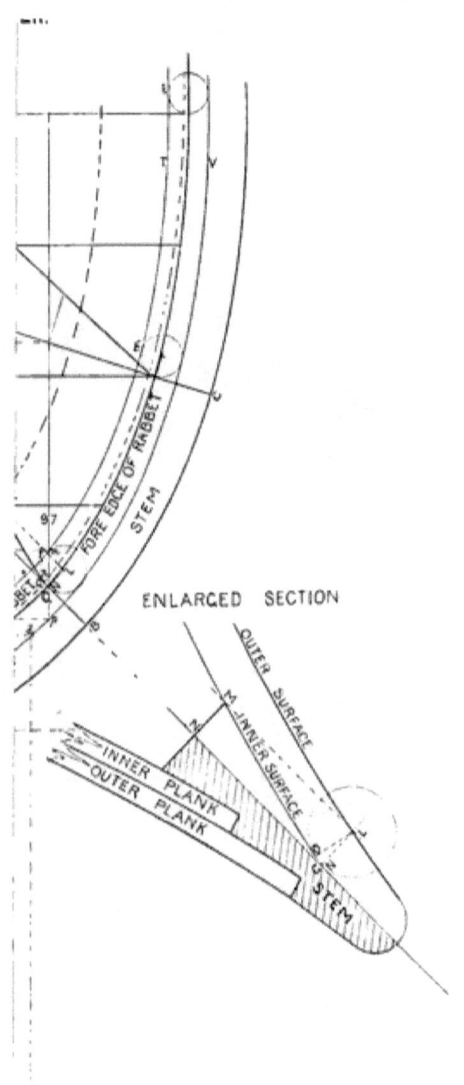

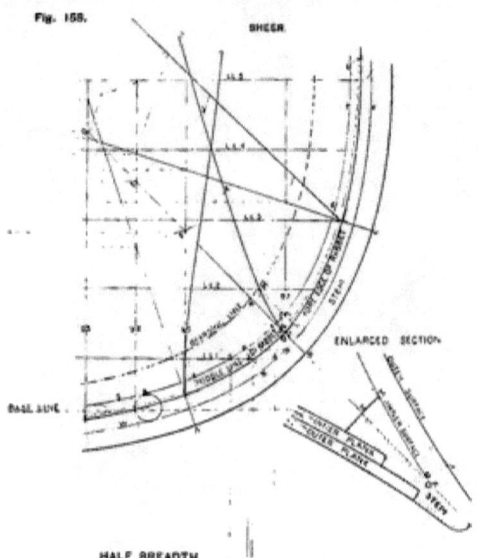

Fig. 158.

HALF BREADTH

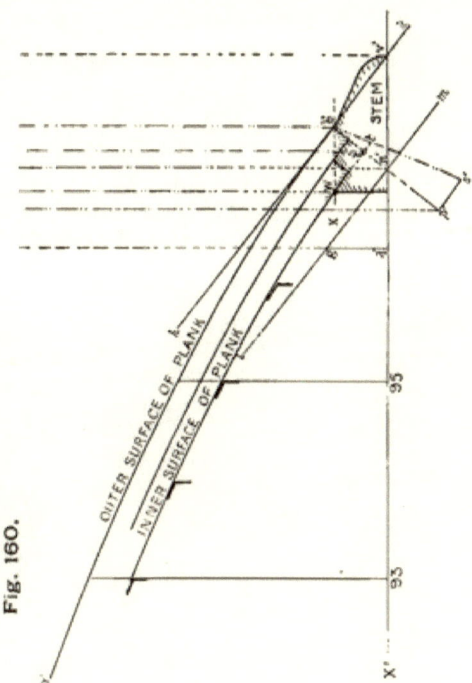

Fig. 160.

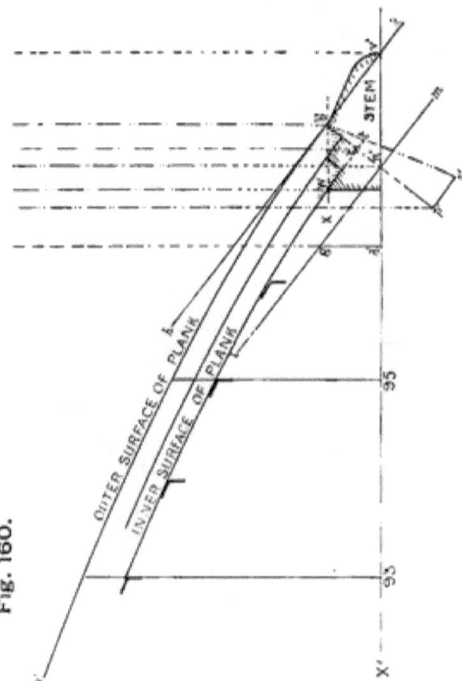

Fig. 160.

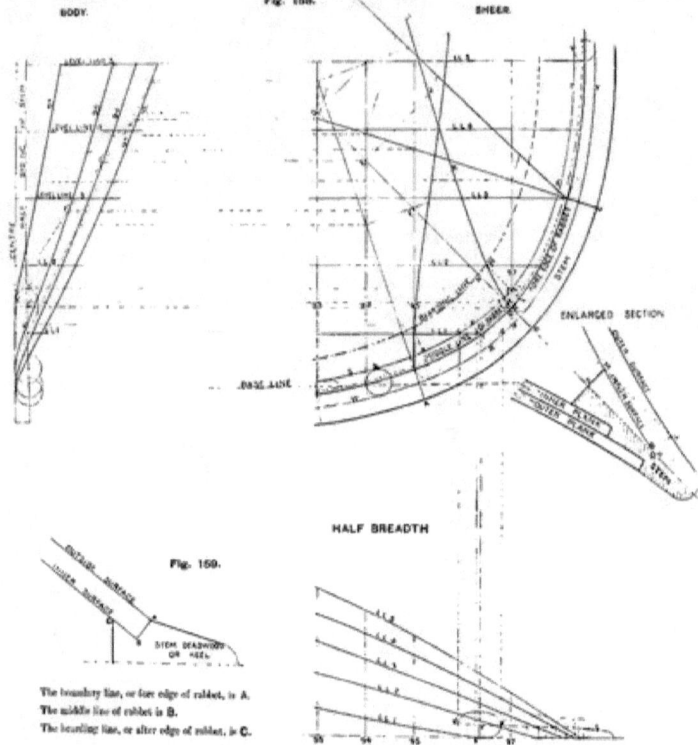

The boundary line, or fore edge of rabbet, is A.
The middle line of rabbet is B.
The bearding line, or after edge of rabbet, is C.

in r^1, which point project upon the level line in the Sheer at r. Also through the point g, draw $l\,m$ parallel to $h\,k$, cutting the middle line at n^1, which point project upon the base line of the Sheer at n. Join $n\,r$, then the line $n\,r$ is the vertical trace of the tangent plane to the surface of the ship at b, b^1. Through b and b^1 draw $b\,p$ and $b^1\,p^1$ perpendicular to $n\,r$ and $n^1\,l$ respectively, that is, to the vertical and horizontal traces of the tangent plane. Therefore $b\,p$ and $b^1\,p^1$ are the vertical and horizontal projections of a perpendicular to the tangent plane at the point b, b^1. Take any point, p, p^1, in the perpendicular, and consider for the present that the plane of the level line $a\,b$ is the horizontal plane of projection. Through p^1 draw $p^1\,s^1$ perpendicular to $p^1\,b^1$, and make $p^1\,s^1 =$ to $p\,s$ in the Sheer. Join $b^1\,s^1$, this is the rebatment on the horizontal plane of a line perpendicular to the tangent plane at b, b^1. Set off on $b^1\,s^1$ the distance $b^1\,t$ equal to the thickness of the bottom plank, and through t draw $t\,t^1$, parallel to $s^1\,p^1$, cutting $b^1\,p^1$ at t^1, then t^1 is the *horizontal* projection of a point in *the middle of the rabbet*. By projecting the point t^1 upon the line $b\,p$ a point t^2 is found in the vertical projection of the middle of rabbet. Similarly, the point t^2 may be projected over the Body at t^3. Other points in the middle of rabbet having been found in the three plans, the line of the middle of rabbet can be drawn, and all the level lines in the Sheer and Body will end at their intersections with it. These endings can then be projected into the Half Breadth, and the level lines in that plan ended at the points given thus.

To Find the Bearding Line by another Method. – In Fig. 160 draw in the Half Breadth short lines parallel to the middle line of that plan b^1 W, and at a distance from it equal to the half-siding of the stem at each of the level lines, the points where these lines intersect the corresponding level lines, of the *inner* surface of the plank, will be in the horizontal projection of the bearding line, and the vertical projection of that line is found by projecting these intersections W upon the corresponding level lines W^1 in the Sheer, and passing a curve through the points W^1, etc., so obtained.

Form of the Rabbet in the Main Keel Piece. — In Fig. 161 from the outer boundary point A describe circle C J B equal in radius to the thickness of the plank at that part, which is usually slightly thicker than the other planking. Produce the inner surface of the planking L N tangent to the circle cutting at C. Join C to A. Continue E A to J, and draw J M level. Then J M is the top of the

in r^1, which point project upon the level line in the Sheer at r. Also through the point g, draw $l\,m$ parallel to $h\,k$, cutting the middle line at n^1, which point project upon the base line of the Sheer at n. Join $n\,r$, then the line $n\,r$ is the vertical trace of the tangent plane to the surface of the ship at b, b^1. Through b and b^1 draw $b\,p$ and $b^1 p^1$ perpendicular to $n\,r$ and $n^1 l$ respectively, that is, to the vertical and horizontal traces of the tangent plane. Therefore $b\,p$ and $b^1\,p^1$ are the vertical and horizontal projections of a perpendicular to the tangent plane at the point b, b^1. Take any point, p, p^1, in the perpendicular, and consider for the present that the plane of the level line $a\,b$ is the horizontal plane of projection. Through p^1 draw $p^1\,s^1$ perpendicular to $p^1 b^1$, and make $p^1 s^1 =$ to $p\,s$ in the Sheer. Join $b^1 s^1$, this is the rebatment on the horizontal plane of a line perpendicular to the tangent plane at b, b^1. Set off on $b^1 s^1$ the distance $b^1 t$ equal to the thickness of the bottom plank, and through t draw $t\,t^1$, parallel to $s^1 p^1$, cutting $b^1 p^1$ at t^1, then t^1 is the *horizontal* projection of a point in *the middle of the rabbet*. By projecting the point t^1 upon the line $b\,p$ a point t^2 is found in the vertical projection of the middle of rabbet. Similarly, the point t^2 may be projected over the Body at t^3. Other points in the middle of rabbet having been found in the three plans, the line of the middle of rabbet can be drawn, and all the level lines in the Sheer and Body will end at their intersections with it. These endings can then be projected into the Half Breadth, and the level lines in that plan ended at the points given thus.

To Find the Bearding Line by another Method. — In Fig. 160 draw in the Half Breadth short lines parallel to the middle line of that plan b^1 W, and at a distance from it equal to the half-siding of the stem at each of the level lines, the points where these lines intersect the corresponding level lines, of the *inner* surface of the plank, will be in the horizontal projection of the bearding line, and the vertical projection of that line is found by projecting these intersections W upon the corresponding level lines W^1 in the Sheer, and passing a curve through the points W^1, etc., so obtained.

Form of the Rabbet in the Main Keel Piece. — In Fig. 161 from the outer boundary point A describe circle C J B equal in radius to the thickness of the plank at that part, which is usually slightly thicker than the other planking. Produce the inner surface of the planking L N tangent to the circle cutting at C. Join C to A. Continue E A to J, and draw J M level. Then J M is the top of the

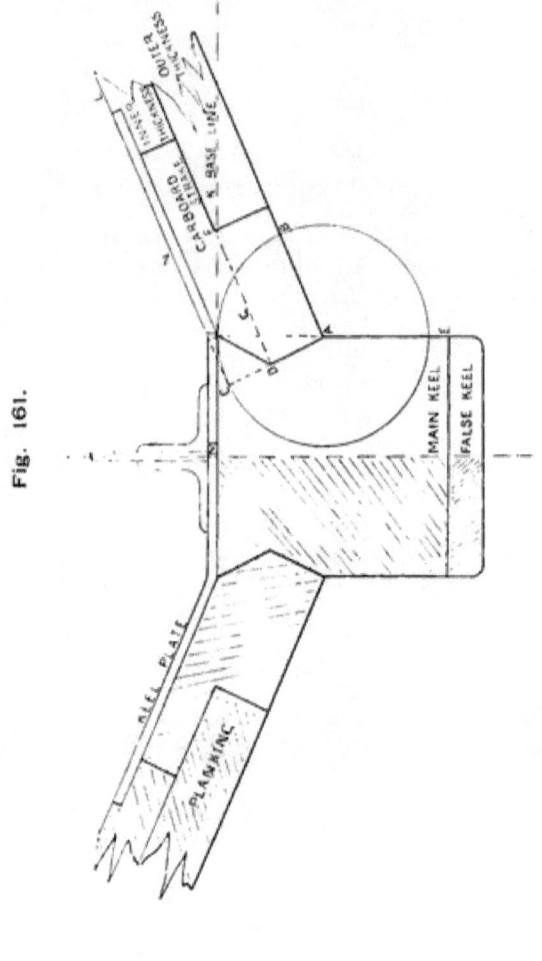

Fig. 161.

main keel piece. Continue inner thickness line F G to cut A C. Then D is the middle of the rabbet. Join D to J, which gives A D J the form of the rabbet for the thick garboard strake.

Working Base Line.—Level J extreme height of the circle C J B on to the centre line, then M K is the position of the conventional base line for laying off, which is produced fore and aft on the loft floor.

SHEATHED VESSELS.

CHAPTER XX.

Alteration of Practice—Thickness of Wood Sheathing—Thickness of Shell Plating—Method of Fastening—Solid Stems and Stern Posts—Hollow Section Stems and Stern Posts—Method of Housing, Planking and Shell Plating—Connection of Keel—Finish of Planking and Plating on the Stern—Stern Post of a Cruiser—Stem of a Cruiser—Shaft Brackets or Struts—Method of Attachment at Head and Foot—Necessary Moulds—Stern Tubes in Twin Screws—How to House Planking and Shell Plating—Bossing of Frames—Necessary Moulds—Stern Posts in Single Screw Vessels—System of Terminating Planking, Shell Plating, and Keel—Necessary Moulds for Casting—Rabbet in the Main Keel Piece—Taking Off the Planking and Shell Plating.

The practice of double planking, adopted in composite vessels, for so many years by the British and other navies, is giving place to single planking of $3\frac{1}{4}$ to 4 inches, placed on top of a steel shell of at least a $\frac{1}{4}$ of an inch thick. This planking, which is formed of 12 inch strakes, is attached to the plating between the frames by $\frac{7}{8}$ of an inch forged naval brass bolts, tapped through the plating, and having a grummet and steel washer under the nut of the bolt on the inside. The lap edges of the plates are generally made single, and the butts double riveted. This planking extends above the load-line amidships for about 2 feet, and at the forward and after ends 5 and 3 feet respectively. Solid stern posts and stems of gun-metal or phosphor bronze have been fitted, but these are giving place to hollow sections, as shown by **Figs. 162** and **163**, made of gun-metal. In settling the size of the section, care should be exercised on the loft floor to keep the moulded line produced about 1 inch from the centre line at the fore edge of the posts to allow room for riveting shell to the casting. The shell plating is rabbeted on to the post as shown, and provision is made for housing the planking, so that it is flush with the outside of the casting. The keel piece of the post has a notch for receiving the flat plate keel, and a shoe A for securely attaching the main keel piece. **Fig. 162** is the stern post of a cruiser with twin screws, and a recess is provided in the sole-piece for

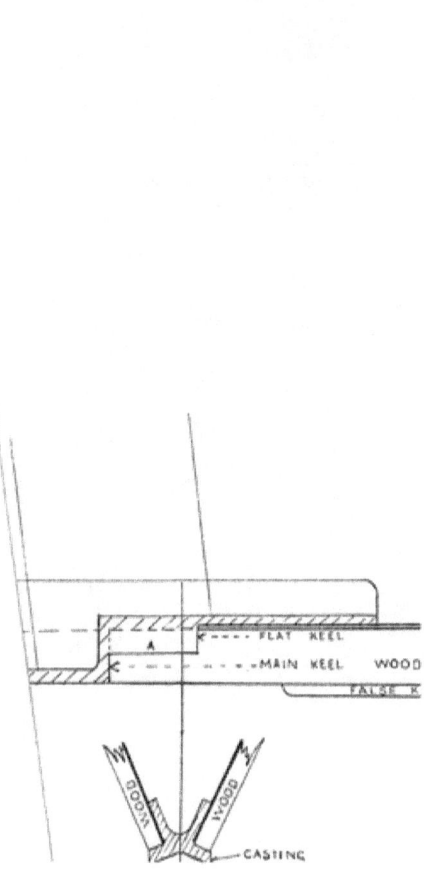

Fig. 162.

STERN-POST OF A SHEATHED CRUISER.

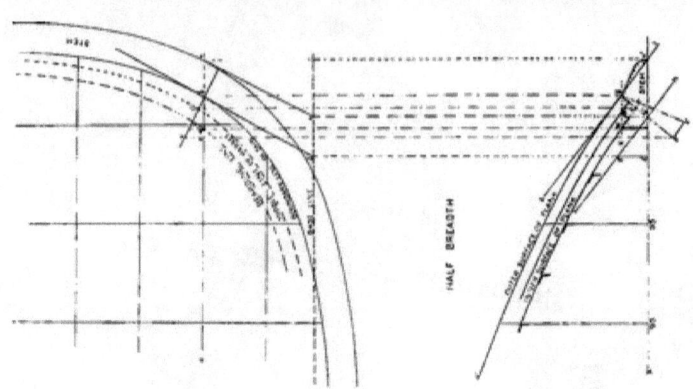

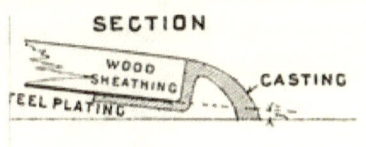

SECTION

SECTION

ECTION

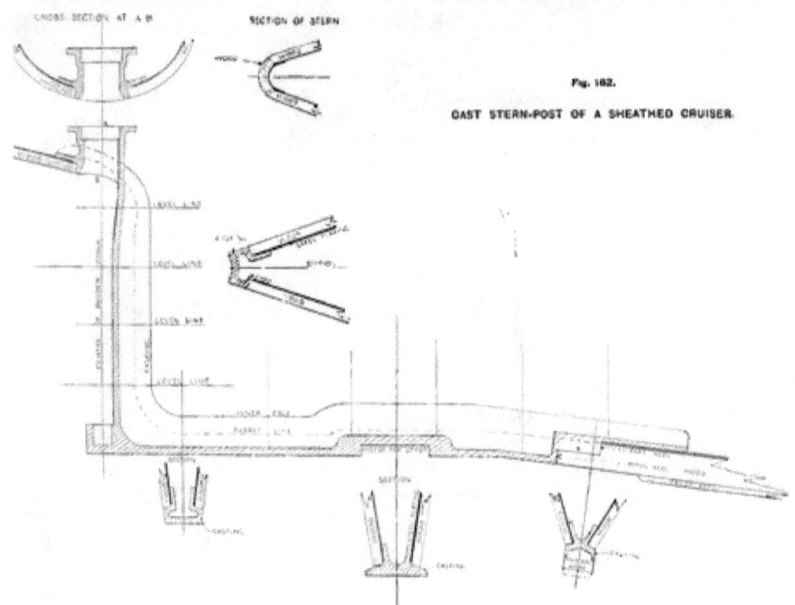

Fig. 162.
CAST STERN-POST OF A SHEATHED CRUISER.

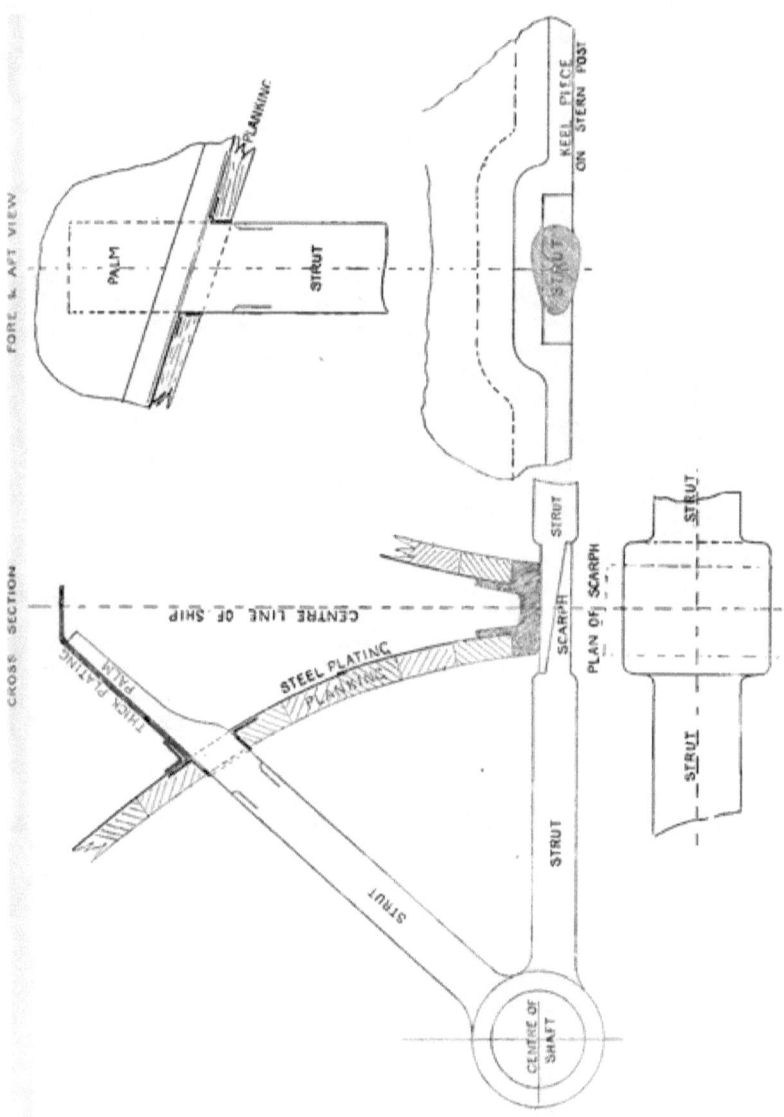

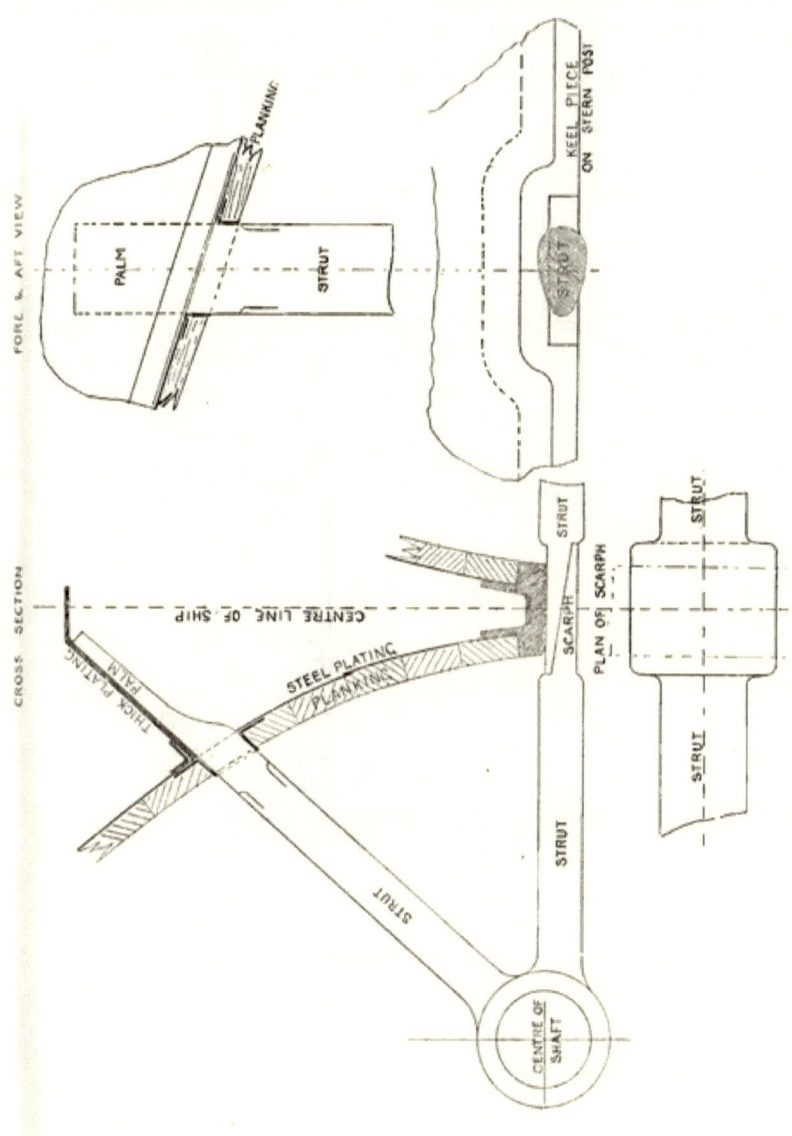

receiving the after brackets or struts, fitted for the support of the end of the screw shafts, the connection of which will be seen by reference to **Fig. 164**.

By one of the detail sections of **Fig. 162** it will be noted that a vertical centre plank is worked up the stern plate from the stuffing box projection to the top to receive the plank ends.

Shaft Brackets or Struts Aft.—Occasionally these are made in one piece, but a more general practice seems to be to fit them, as shown by **Fig. 164**, with a V scarph over the keel. This method allows easy adjusting for the shaft centres. The upper palms are attached to the protective deck, or a strong saddle-back between

Fig. 165.

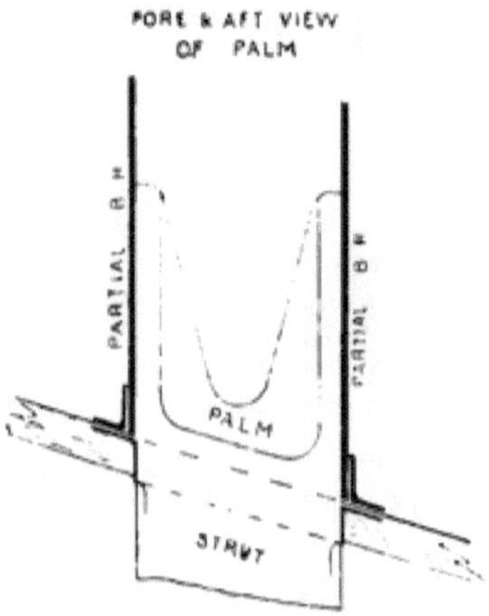

partial bulkheads. They may, where found necessary, be fitted with a double palm as in **Fig. 165**. A watertight collar, in two pieces, is worked on the shell plating round the struts; the flange being made sufficiently deep to take the wood sheathing as shown.

To make the patterns for casting, the loftsman supplies a drawing, on loose boards, of a vertical cross section through the centre of the strut, with the centre of shaft and fore and aft bevels for the palms. A small detailed figured drawing is given with the mould, which enables patterns to be made.

Stern Tubes in Twin Screw Vessels.—These tubes are made of gun-metal, as in **Fig. 166**, shown by sectioned parts. Provision is

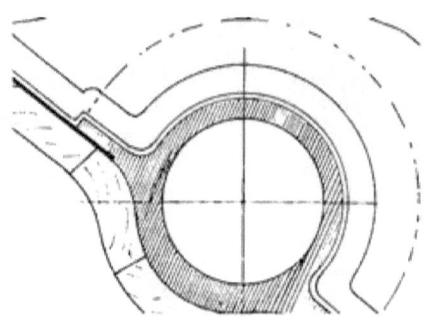

Fig. 163.

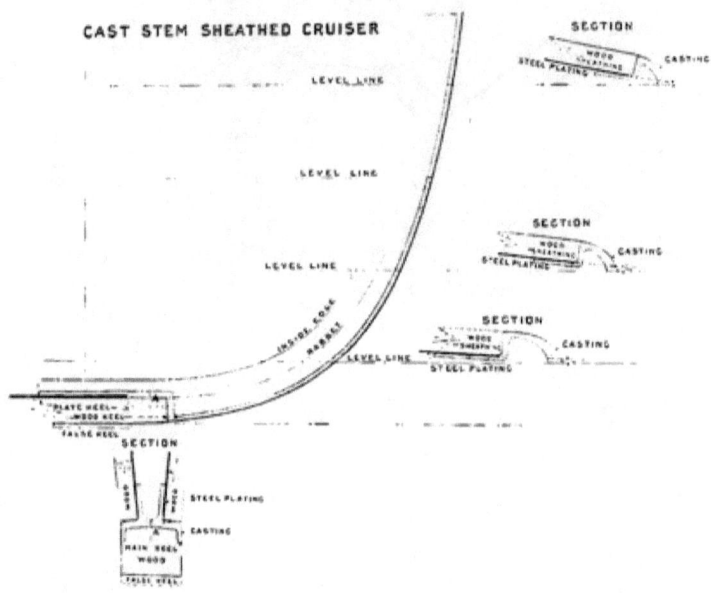

SHEATHED VESSELS. 163

made for housing the wood planking and shell plating with watertight connection. They are attached by projecting palms to partial bulkheads, shown at A B, F E, and J K sections. The intermediate frames between H and K are bossed in, as seen in section G H, and a connection to tube made at P. Those forward of K J are bossed out sufficiently to allow the withdrawing of the engineer's tube, etc., or one or two frames are cut to save excessive bossing, and fore and afters fitted to maintain rigidity. The after end has a projecting flange L, with a notch for taking shell plating and housing planking :

Fig. 169.

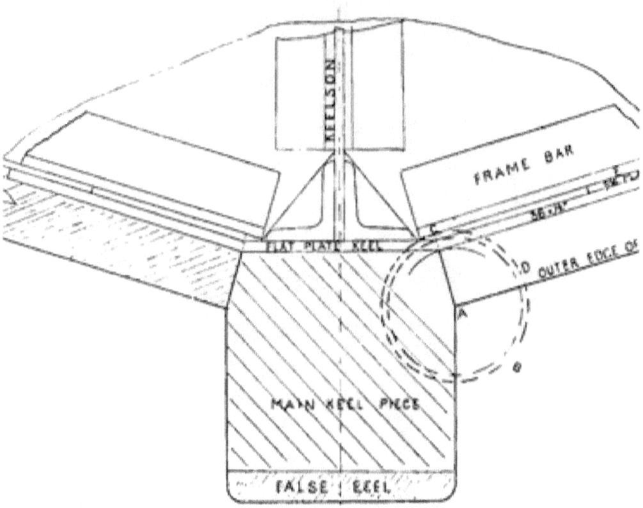

this is carried round the full length of the tubes as indicated by the sections at different points. The form of the frames in way of the tubes are scrieved in on the boards, after fairing up on the loft floor.

The loftsman supplies, on a board, the pattern-makers with a longitudinal section of the sectioned parts, together with the cross sections, as represented, having marked on them vertical and horizontal winding lines to get the correct bevel of the ship's side. The pattern is usually carefully checked by the loftsman before being sent away.

SHEATHED VESSELS. 163

made for housing the wood planking and shell plating with watertight connection. They are attached by projecting palms to partial bulk-heads, shown at A B, F E, and J K sections. The intermediate frames between H and K are bossed in, as seen in section G H, and a connection to tube made at P. Those forward of K J are bossed out sufficiently to allow the withdrawing of the engineer's tube, etc., or one or two frames are cut to save excessive bossing, and fore and afters fitted to maintain rigidity. The after end has a projecting flange L, with a notch for taking shell plating and housing planking :

Fig. 169.

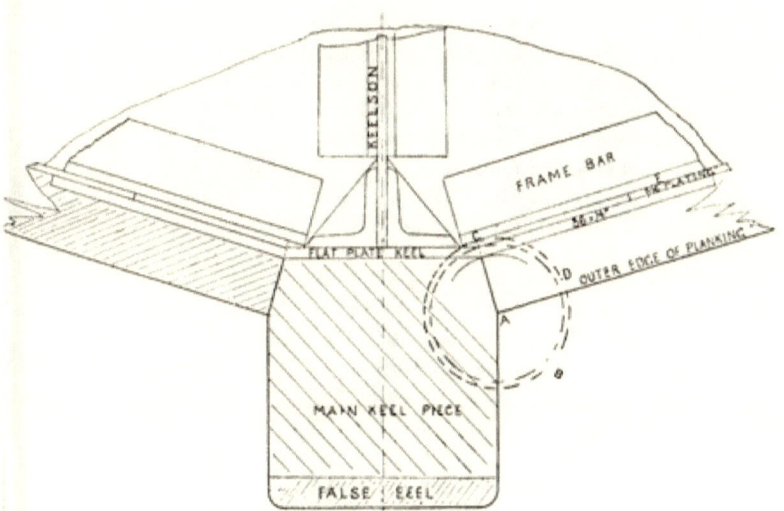

this is carried round the full length of the tubes as indicated by the sections at different points. The form of the frames in way of the tubes are scrieved in on the boards, after fairing up on the loft floor.

The loftsman supplies, on a board, the pattern-makers with a longitudinal section of the sectioned parts, together with the cross sections, as represented, having marked on them vertical and horizontal winding lines to get the correct bevel of the ship's side. The pattern is usually carefully checked by the loftsman before being sent away.

Stern Posts in Single-screw Vessels.—They have been fitted in two pieces, as in **Fig. 167**, the sole-piece being keyed at C into the stern post—shown in section through the key-way—and scarphed at D. The head of the post is attached to a strong horizontal plate by a vertical **T** bracket. The rudder post, of wood, is fitted into a hollow piece cast on the sole-piece. The planking and shell plating is rabbeted into the head of the stern post. Provision is made for securing the planking and shell plating, as shown by section through A B. The fore end of the sole-piece is made with a shoe and notch to take the main wood and flat plate keel.

A better method is that given in **Fig. 168**, where the brass casting is in one piece. The sections show clearly the character of the post, and the way of terminating the shell plating and planking. For making the patterns a full-sized drawing, as shown, is given by the loft, with sections at different points of the keel piece, and others at the level lines, a few through the arch over the aperture, and about three cross sections through the head piece.

Rabbet in the Main Keel Piece.—Referring to the **Fig. 169**. From the point A, with A D radius, equal to the thickness of the planking and the garboard shell plating, describe circle B D C. Produce moulded line E on the heel of the frame, tangent to this circle, cutting say at C, join A to C. Line in the flat plate keel as shown, then A C is the line of the rabbet for terminating the planking.

Taking off the Planking, etc.—The planking and shell plating are taken off in the same manner as explained on page 154 for composite vessels.

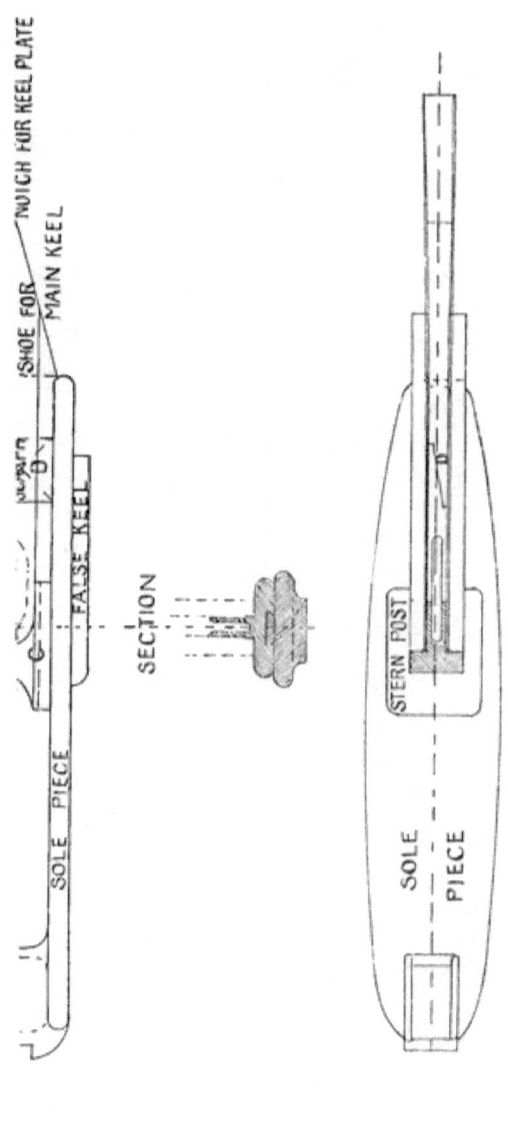

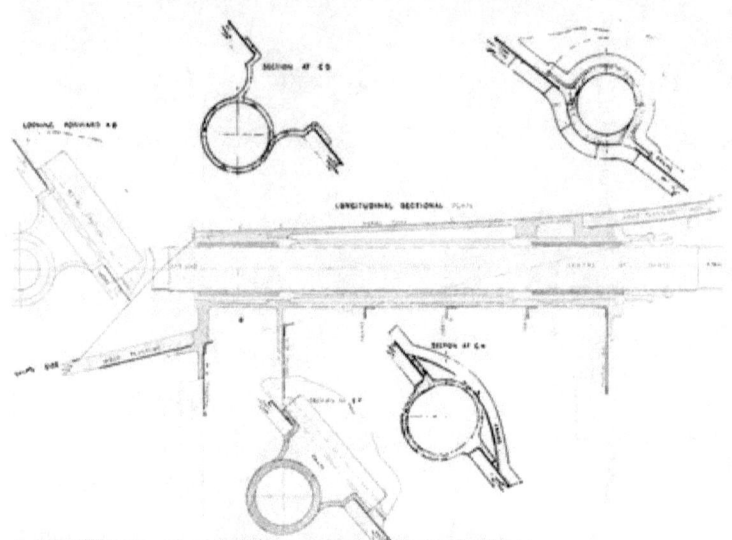

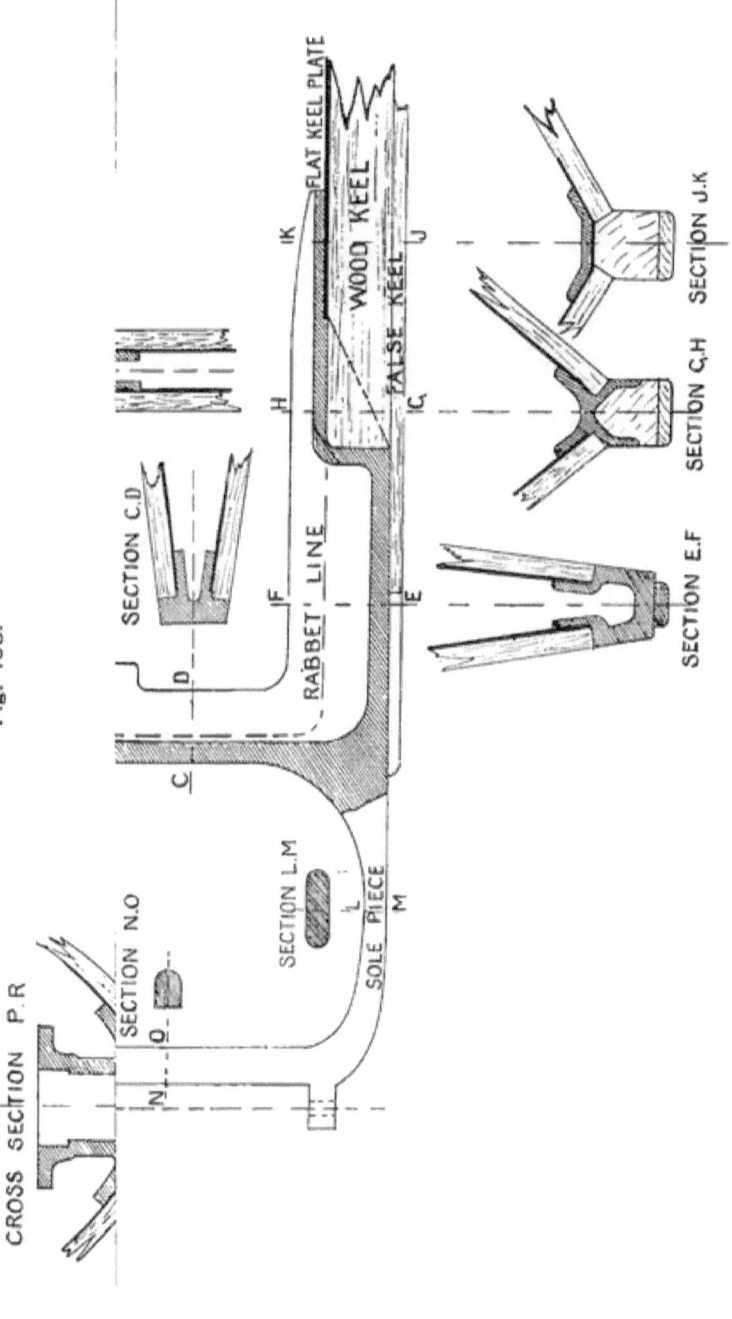

Fig. 168.

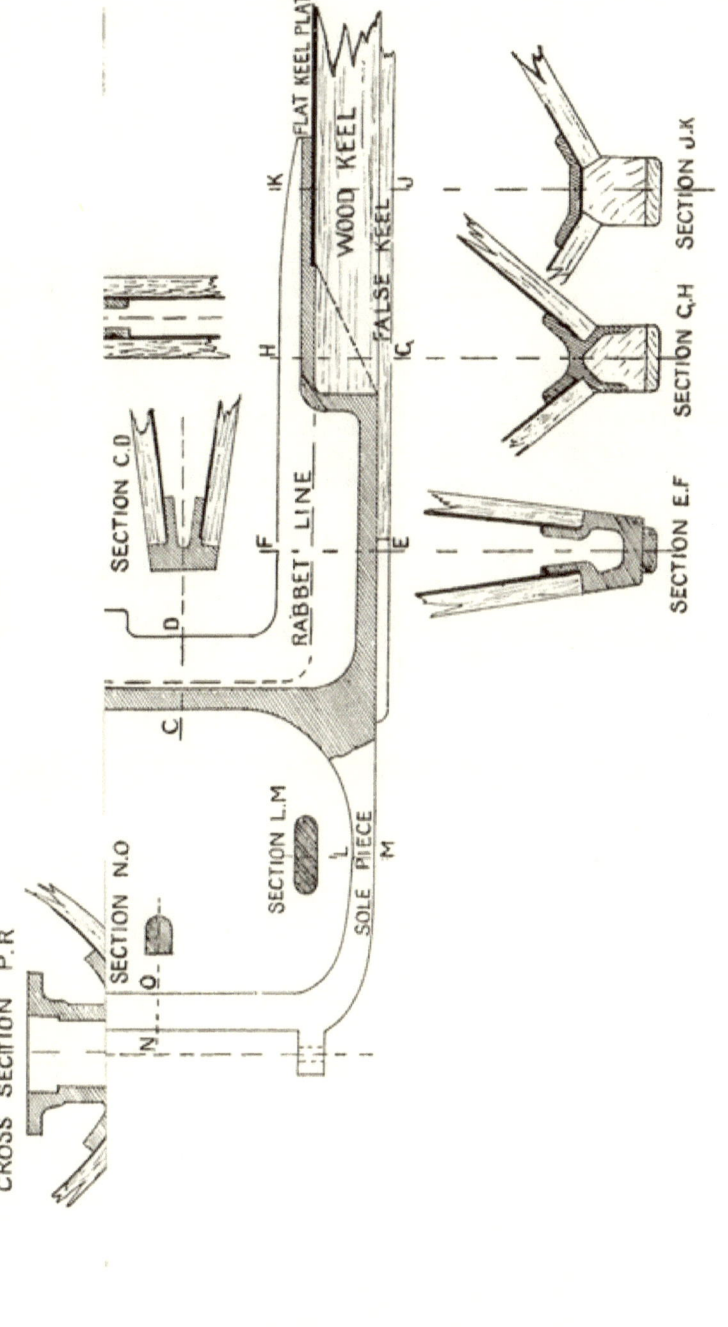

Fig. 168.

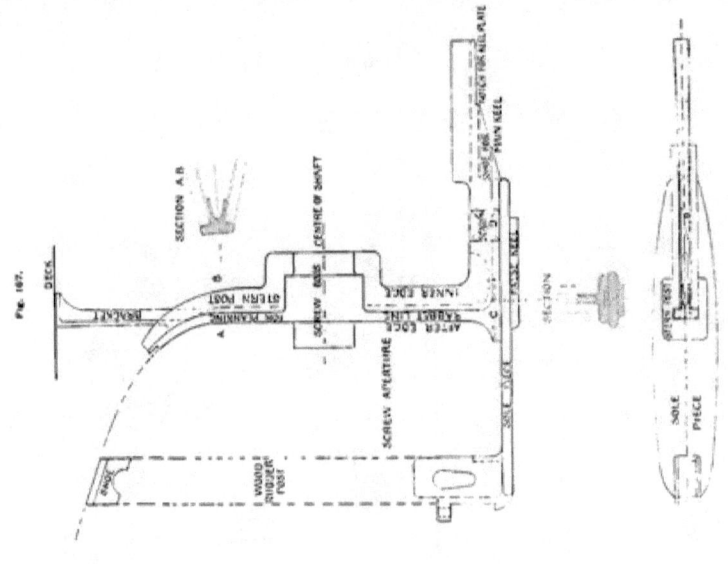

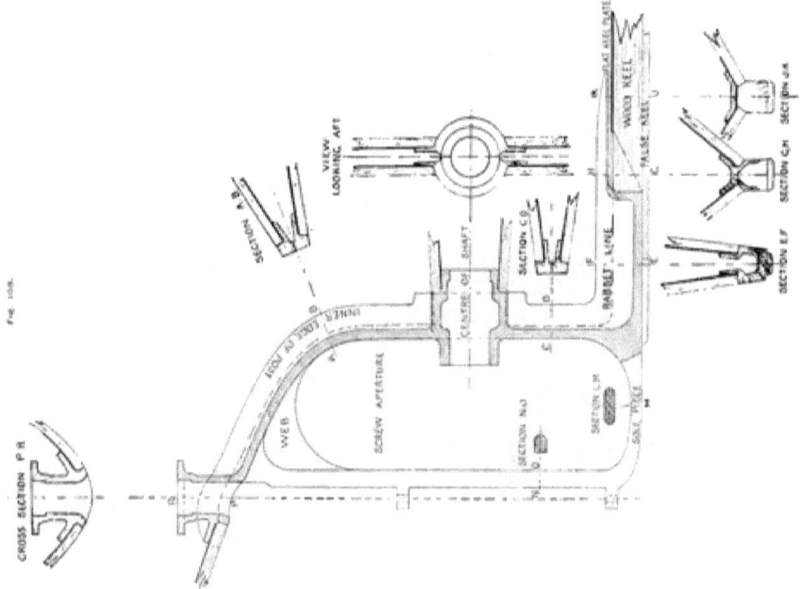

INDEX.

After bossed frames, 12, 14
  ,, end finish, 16
  ,, end floors, 38
  ,, end supports for shafts, 77, 149
Allowance for beam arms, 88
Alteration of practice in sheathed ships, 160
Appearance of lines in sheer draught, 2
Applying frame bevels, 63
Approximate sheer line, 20
Area of inner bottom plating, 42
  ,, outer bottom plating, 49
Armour barbette, 124
  ,, belt, battle ship, 119
  ,, belt connection, 120
  ,, belt framing, 119
  ,, conning tower, 149
  ,, in battle ship, 119
  ,, in cruiser, 114
  ,, shelf, 119
  ,, turret, 126

Ballast tank, cellular bottom, 41
  ,, tank, McIntyre type, 44
  ,, tank, Swan type, 44
Barbette of circular form, 126
  ,, of egg form, 124
  ,, structure, 124
Base line in sheer draught, 7, 159
Battle ship's belt deck, 119
  ,, inner bottom, 137
  ,, protective deck, 121
  ,, side armour, 119
Beam arm allowance, 88
  ,, bevels, war vessels, 118
  ,, camber, merchant ships, 6, 34
  ,, camber, war vessels, 149
  ,, moulded, 3

Beams protective deck, 118
Bearding line, approximate, composite vessels, 156
Bearding line, accurately, composite vessels, 157
Bell-mouthed cargo hatches, 102
Belt armour battle ship, 119
  ,, correction, 119
  ,, expansion, 123
  ,, fairing, 123
  ,, moulds, 124
Belt deck construction, 119
Best lines for fairing-up, 8
  ,, position for bilge keels, 128
Bevelling board for harpins, 70
Bevelling machine, 63
Bevels for frames, 62
  ,, inner bottom frames, 139
  ,, war ship's beams, 118
Bilge diagonal, 8
  ,, keels, 128
  ,, turn, 6
Boat beam moulds, 84
  ,, davits and chocks, war vessels, 151
Body plan explained, 2
  ,, fairing-up, 7
Bossed frames, aft, 12, 14
  ,, at ram, 129
  ,, sheathed ships, 163
Bossing twin screws, 12, 148, 163
Bottom plating area, 42, 48, 132
  ,, rise, 5
Bow and buttock lines, 11
Box framing, battle ships, 121
Brackets for struts, 77, 149, 162
Bridge mould, 151
Butts of belt deck, 121
  ,, protective deck, 118

Calculation of shaft passage plating, 104
Camber of beam allowed, 34
,, in war vessels, 149
,, methods of laying off, 34
Cant frames aft, 32
,, forward, 27
,, ,, expanded, 27
,, moulds, 33
Cant knees on scrieve board, 58
Cargo hatches, bell-mouth type, 102
,, ordinary type, 101
Cellular double bottom, 41
Centre through plate keelson mould, 83
Change of frames, 6
Check on stern expansion, 51
Checking frame bevels, 63
,, keel, 110
Chocks for boats in war vessels, 151
Circular barbette, 126
Classification depth, 6
,, length, 6
Clipper type of stem, 107
Common harpin, 69
Composite vessels, 154
Conical ballast tank, Swan's, 41
,, gun gallery, 142
Connection of belt deck, 119
,, sheathed ship's keels, 163
Conning tower, 149
Construction of belt deck, 119
,, sheer diagram, 22
Contracted method of fairing, 10
Correct method of protective deck expansion, 117
Correction for belt armour, 119
Cruiser's armour, 114
,, protective deck, 114
Curved floors, 37
Cutting hawse pipe holes, 102
Cut-water, 107

Davits for war ship's boats, 151
Deadflat, 5
Deck houses of iron, 100
,, lines on scrieve board, 56

Deck plate edges, 36
,, ribbands, 68
,, sheer at side, 35
,, stringer plate expansion, 35, 88
,, surface-expanded, 35
,, wide stringer plates, 36
Definition of bilge diagonal, 11
,, bow line and buttock, 11
,, fair line, 9
,, fairness, 9
Depth by classification, 6
,, moulded, 5
,, of hold, 6
Detective bevelling machine, 63
Diagonal cutting knuckle, 20
Diagonals, 8
Diagrams for sheer line, 22
Diminishing line of floors, 37
Double bottom expansion, 42
,, ,, floors, 44
,, ,, frame bevels, 139
,, ,, margin plate, 43
,, bottoms, 41, 130
,, cellular bottom, 41
,, McIntyre bottom, 44
,, Swan bottom, 44
Doubling of masts at deck, 92
,, ,, heel, 92
Draft marks, merchant, 108
,, ,, war vessel, 152
Drawing in deck at side, 35
,, sheer-draught, 7

Edges of belt deck, 121
,, deck plating, 36
,, outer bottom plating, 45
Ellipse, 113
End armour, battle ship, 121, 122
,, floors, 38
,, of level line in half breadth, 8, 156
Exact method of finding heel of frames in composite vessels, 155
Expansion of armour belt, 123
,, barbette armour, 124
,, battle ship's inner bottom, 139

INDEX.

Expansion of cargo hatches, 101, 102
,, conical gun gallery, 143
,, deck stringer plate, 88
,, deck surface, 35
,, double bottom floors, 44
,, end gun gallery, 145
,, forward cants, 27
,, inner bottom. cruiser, 132
,, inner bottom plating, 42, 139
,, iron deck house, 100
,, longitudinals, cruiser, 132
,, longitudinals on curved diagonals, 133
,, margin plate, 43
,, mast plating, 91
,, midship gun gallery, 141
,, poop round, 86
,, protective deck, 115
,, rudder trunk, 96
,, shaft tunnel, 104
,, sheer harpin, 70
,, shell, 48, 49
,, shell plate, 50
,, stern, 51
,, stern of ordinary type, 51
,, thin plates of barbette, 124
,, tumble-home stern, 53
,, turned up floors, 38
,, turtle back, 87
Explanation of body plan, 2
,, half breadth plan, 2
,, profile, 2
,, sheer draught, 1
,, terms used, 3
Extra length to form beam arms, 88
Extreme end floors, 38, 44
,, form, composite vessels, 154

Fairing by contraction, 10
,, double bottom, 41, 130
,, double bottom, battle ship, 137
,, up belt armour, 123

Fairing up body plan, 7
,, conical gun gallery, 142
,, end gun gallery, 145
,, floors, 38
,, frames by diagonal, 20
,, half breadth, 8
,, midship gun gallery, 141
,, on loft floor, 7
,, protective deck, 115
,, semi-egg formed gallery, 144
,, shell plating edges, 47, 48
,, stern, 18
,, turtle back, 87
Fair line, 9
Fairness in sheer draught, 9
Fall in of bilge, 5
Figure head lacing piece, 108
,, ,, moulds, 108
,, step, 107
Final test of fairness, 10
Finding at ship moulded depth, 106
,, heel of frames, composite vessels, 154
Finish of after end, 16
,, belt armour ends, 123
,, planking, sheathed ship, 160
Flam, 6
Flat plate keel mould 79
Floors at extreme ends, 38
,, in double bottom, 44
,, on scrieve board, 58
,, turned up, 37
Forecastle head, 108
Forefoot, 6
Form of bell-mouth cargo hatches, 102
,, common harpin, 69
,, forward cants in sheer, 27
,, iron deck-house, 100
,, lines in sheer draught, 2
,, ordinary cargo hatches, 101
,, rabbet, keel piece, 157, 164
,, ribband line, 67
,, sheer harpin, 70
,, stern cants in sheer, 32
,, stern harpin, 72
Forward cant frames, 27

Forward gun galleries, 144
Frame bevels, 62, 139
  ,, ,, application, 63
  ,, ,, checked, 63
  ,, ,, inner bottom, 139
  ,, bossed aft, 12, 14
  ,, ,, forward, ram, 129
  ,, ,, sheathed ships, 163
Frame bevel tester, 63
Frames canted in fore body, 27
  ,, change, 6
Framing behind side armour, 119
Freeboard screw steamer, 104
  ,, sailing ship, 105

General description of box-framing, 121
General description of barbette structure, 124
General terms used, 3
Gun galleries, 141

Half Breadth fairing-up, 8
  ,, plan explained, 2
Handy levelling machine, 64
Harpin, bevelling board, 70
Harpins, 67
Hatches for shipping cargo, 101, 102
Hawse pipes, 102
Head of forecastle, 108
Heel of frames, composite vessels, 154
Hollow section stems and stern posts, 160
How lines appear, sheer-draught, 2
  ,, scrieve board is prepared, 54
  ,, to house planking, etc., sheathed vessels, 160
  ,, to lift deck sheer side, 56
  ,, to obtain mast lines, 91
  ,, ,, round poop lines, 85
  ,, ,, turtle back lines, 87

Information on scrieve board, 54
Inner bottom, battle ship, 137
  ,, cruiser, 130
  ,, expansion, 42, 132, 139
  ,, frame bevels, 139
  ,, longitudinal expansion, 133, 135
  ,, longitudinals, 132
  ,, ribbands, 68
Iron deck house, 100
Iron and steel masts, 91

Keelsons on scrieve board, 58
  ,, moulds, 83
Keel plate mould, 79
  ,, rabbet, composite vessel, 157
  ,, ,, sheathed vessel, 164
  ,, scrieve boards, 81
Keels on bilge, 128
Knees for beams, 88
  ,, cants, scrieve board, 58
  ,, tank side, scrieve board, 44
Knuckle cut by diagonal, 20

Lacing-piece figure head, 108
Laying-off beam camber, 34
  ,, on diagonals, 9, 55
  ,, on left floor, 6
  ,, ribbands, 67
Length between perpendiculars, 3
  ,, by classification, 6
  ,, over all, 3
Level line ending, composite ship, 156
Lifting beams, 59
  ,, bevels of cants, 33
  ,, ,, frames, 62
Line for base, 7, 159
  ,, off draft marks, 108, 152
  ,, of sheer, 20
Lines composing sheer draught, 2
  ,, for decks, scrieve board, 56
  ,, for fairing-up, 8

INDEX.

Lining off model, 46
 ,, shell on model, 46
Longitudinals of inner bottom, 42, 130
Lowest point of sheer, 23

Machine bevelling, 65
 ,, for frame bevel testing, 63
Margin plate expansion, 43
Marking off freeboard, 104
 ,, hawse pipes, 102
Mast doubling, 92
 ,, expansion, 91
 ,, form, 91
 ,, tube expansion, 92
McIntyre ballast tank, 44
Method of attaching struts, sheathed ships, 162
 ,, fastening, sheathed ships, 160
 ,, finding oval form, 113
 ,, ,, sheer line, 20, 35
 ,, housing planking, etc., sheathed ships, 160
 ,, laying-off at Swan & Hunter's, 25
 ,, ,, beam camber, 34
 ,, ,, cants, 27, 32
 ,, ,, stringer plate, 88
Middle line of rabbet, composite vessels, 155, 156
Midship gun gallery, 141, 142
 ,, section, 3
Miscellaneous, 96.
Mocking-up system expansion, 135
Mode of laying-off, 6
 ,, plating protective deck, 117
Model of protective deck, 118
More correct protective deck expansion, 117
Moulded beam, 3
 ,, depth, 5
 ,, depth at ship, 106
 ,, form, composite vessels, 154
Moulds for barbette armour, 124
 ,, beams, 34, 59, 149
 ,, belt armour, 124

Moulds for boat beams, 84
 ,, boat davits, 151
 ,, cant frames, 33
 ,, carver, 108
 ,, centre through plate keelson, 83
 ,, flat plate keel, 70
 ,, pilot bridge, 151
 ,, stem, 74, 148, 160
 ,, stern frame, 77, 147, 160
 ,, stern struts, 77, 149, 162
 ,, stern tubes, 78, 148, 162
 ,, necessary, merchant vessels, 74
 ,, ,, war vessels, 147

Necessary moulds, 74, 147

Obtaining true form of shell plating, 49
Obtaining barbette armour, 126
 ,, conical gun gallery, 142
 ,, double bottom, battle ship, 137
 ,, double bottom, cruiser, 130
 ,, double bottom, merchant ship, 41
 ,, end gun gallery, 145
 ,, form, rudder trunk, 96
 ,, ,, shaft tunnel, 103
 ,, midship gun gallery, 141
 ,, protective deck form, 114
 ,, ,, ,, expansion, 115
 ,, semi egg-shaped gun gallery, 144
 ,, shell expansion, 48
 ,, shell sight edges, 45, 127
 ,, tank knees, 44
 ,, turtle back lines, 87
Ordinary cargo hatches, 101
Ordering shell plating, 50
Outer bottom area, 48
 ,, ,, edges, battle ship, 127
 ,, ,, ,, cruiser, 128
 ,, ,, ,, merchant ship, 45
Oval form, 113

13

INDEX.

Particulars to lay ship down, 3
Pilot bridge, 151
Placing shell plating edges on model, 46
Planking finish on stem, sheathed ships, 160
  ,, finish on stern, sheathed ships, 160
  ,, thickness, sheathed ships, 160
Plate edges of decks, 36
  ,, keel, 79
Plating inner bottom, 42, 130
  ,, of masts, 91
  ,, outer bottom, 45, 127
  ,, protective deck, 117
Poop round expansion, 86
  ,, form, 85
  ,, plating, 85
Preparation of scrieve board, 54
Principal moulds, merchant ships, 74
  ,, war ships, 147
Profile plan explained, 2
Projection in body, forward cants, 29
  ,, in sheer, bilge diagonal, 8
  ,, in sheer, forward cants, 27
  ,, in sheer, stern cants, 32
Protective deck, battle ship, 121
  ,, beams, 118
  ,, correct expansion, 117
  ,, cruiser, 114
  ,, fairing, 115

Rabbet keel piece, composite vessel, 157
Rabbet keel piece, sheathed vessel, 164
Raised keel, 10
Ram, bossed frames, 129
Ready method of finding sheer line, 24
Redoubt, 126
Revolving turret, 126
Ribbands, inner bottom, 68
  ,, form, 67, 68
  ,, scrieve board, 58, 67
  ,, termination aft, 68
  ,, ,, forward, 67
Rise of bottom, 5

Round poop, 85
  ,, up of beams, 6
Rudder trunk, 96

Sailing ship's freeboard, 105
Screw steamer's freeboard, 104
Scrieve board description, 54
  ,, information, 54
  ,, keelsons, 58
  ,, preparation, 54
Scrieve boards for keel, 81
Scrieving in floors, 58
  ,, frames, 55
Seams of protective deck plating, 118
Semi egg-shaped gun gallery, 144
Setting off draught marks, 108, 152
Shaft brackets, 77, 149, 162
  ,, tunnel, 163
Sheathed ships, 160
  ,, bossed frames, 163
Sheer diagram construction, 22
  ,, draught explained, 1
  ,, harpin expansion, 71
  ,, line approximately, 20
  ,, line under freeboard tables, 25
  ,, of deck at side, 20, 35
Shelf armour, battle ship, 123
Shell plating edges, 45, 127
  ,, ,, model, 46
  ,, ,, scrieve board, 48, 57
  ,, expansion, 48, 49
Side armour, battle ship, 119
Sighting the keel, 110
Sny, 45
Solid stems and stern posts, sheathed ships, 160
Sponsons, 141
Stem clipper type, 107
  ,, finish of planking, etc., 160
  ,, mould, merchant ship, 74
  ,, ,, sheathed vessel, 160
  ,, ,, war vessel, 148
  ,, termination for ribbands, 67
Stems and stern posts, hollow section, 160
Step for figure head, 107

Stern cant frames, 32
,, fairing-up, 18
,, finish of planking, 162
,, ,, ribbands, 68
,, frame, merchant vessel, 77
,, ,, sheathed vessel, 160
,, ,, war vessel, 147
,, plating expansion, 51
,, ,, check, 51
,, tubes, merchant vessel, 78
,, ,, sheathed vessel, 162
,, ,, war vessel, 148
Stringer plate expansion, 88
Structure of barbette, 124
Struts, merchant vessel, 77
,, sheathed vessel, 162
,, war vessel, 149
Surface of deck, expanded, 35
Swan and Hunter's method, laying off, 25
Swan's conical tank, 44
System of planking, etc., sheathed vessels, 160

Taking off planking, etc., sheathed vessel, 164
Tank knees, 44
Taper deck stringer plate, 35
Terms used, 3
Test of fairness, 10
The best line for fairing-up, 8
,, common harpin, 69
,, sheer harpin, 70
,, stern harpin, 72
Thickness of planking and shell plating, sheathed vessel, 160
To draw in deck side, 35
,, end level line in half breadth, composite vessel, 156
,, fair frames on diagonals, 20
,, find accurately, bearding line, composite vessel, 157

To find accurately, line of rabbet, composite vessel, 156
,, find approximately, bearding line, composite vessel, 156
,, find heel of frames, composite vessel, 154
,, find middle line of rabbet, composite vessel, 155
,, obtain and line off draft marks, 108, 152
,, obtain form, curved floors, 37
,, ,, inner bottom, battle ship, 137
,, ,, inner bottom, cruiser, 130
,, ,, mercantile double bottom, 41
,, ,, of oval, 113
,, ,, protective deck, 114
,, terminate frame in body, composite vessel, 156
Transferring frame bevels, 63
,, shell sight edges to boards, 57
True form cants, in sheer draught, 27, 32
,, lines, ,, 2
,, of common harpin, 69
Tubes for shaft, 78, 148, 162
Tuck plate, 16
Tumble home, 5
,, ,, stern expansion, 53
Tunnel for shaft, 103
Turned up floors, 37
,, ,, diminishing line, 37
Turn of bilge, 6
Turret, revolving, 126
Turtle back, 87
Twin screw bossing, 12
,, covered in shaft, 12

Use of diagonals, 9

War vessels, 114
Working base line, 7, 159

SOUTH KENSINGTON QUESTIONS
IN
NAVAL ARCHITECTURE.

ELEMENTARY STAGE, 1867 to 1897.

Price, 1s., nett.   By Post, 1s. 0½d.

ADVANCED STAGE, 1878 to 1897.

Price, 1s. 6d., nett.   By Post, 1s. 6½d.

HONOURS STAGE, 1878 to 1897.

Price, 2s. 6d., nett.   By Post, 2s. 6½d.

These Question Books have been found of great value to Students sitting at South Kensington and Board of Trade (Surveyors') Examinations.

NAVAL ARCHITECTS' AND ENGINEERS' DATA BOOK.

ARRANGED FOR COMPILING INFORMATION OF VESSELS.

The book is used by a number of Shipbuilders and others.

Price, 3s. 6d., nett.   By Post, 3s. 8½d.

WATSON'S DOUBLE SLIDING RULE,

FOR CALCULATING DISPLACEMENT, TONNAGE, SPEED, AND INDICATED HORSE-POWER.

This rule is *extremely* easy to understand and work. It is so small in size that it may be carried in an ordinary pocket without inconvenience.

Made of Boxwood, price 12s. 6d., nett.

COMPILED AND SOLD BY

THOMAS H. WATSON,

NAVAL ARCHITECT AND SHIP SURVEYOR.

10, Neville Street, Newcastle-on-Tyne.

www.ingramcontent.com/pod-product-compliance
Lightning Source LLC
Chambersburg PA
CBHW030601300426
44111CB00009B/1062